LIFE

OUT

THERE

LIFE

OUT

THERE

The Truth of
—and Search for—
Extraterrestrial Life

MICHAEL WHITE

THE ECCO PRESS

Copyright © 1998 by Michael White

First published in Great Britain in 1998 by Little, Brown and Company

THE ECCO PRESS
100 West Broad Street
Hopewell, New Jersey 08525

Library of Congress Cataloging-in-Publication Data

White, Michael, 1959–
Life out there : the truth of—and search for—extraterrestrial
life / by Michael White.
p.   cm.
Includes bibliographical references and index.
ISBN 0-88001-671-X
1. Life on other planets. 2. Mars (Planet)—Exploration.
3. Extraterrestrial anthropology. I. Title.
CB54.W49   1999
576.8'39—dc21   98-38377
CIP

9 8 7 6 5 4 3 2 1

FIRST EDITION 1999

For India Grace White, born 9 June 1997

CONTENTS

*Many thanks to Jaimie Tarrell for his advice with
some of the more technical aspects of biology
and to Lisa White for helping with the illustrations.*

INTRODUCTION

How was the Universe created? Why are we here? Does God exist? Are we alone in the Universe?

These are the great questions, the matters that occupy us in the all-too-infrequent moments we can put aside for contemplation, the things we pontificate about after a few drinks. But the harsh reality is that it is probably only the last of these questions that has a fair chance of being answered – sooner or later.

Sadly, if the accountants of the world have their way, the answer to the mystery of whether or not we are alone in the Universe will come later rather than sooner. But, if the spirit of discovery and the determination of scientists prevails, then, maybe, with a great deal of luck, one day – perhaps within our lifetimes – a solution may be found.

In this book I want to address all aspects of the mystery associated with the concept of 'life out there'. I want to present the facts, survey the evidence and to spark off the plethora of questions that any consideration of the subject creates. For example, was there ever life on Mars? If the answer to that is 'yes', then is there still a chance of life being discovered there? How did life originate on Earth? What are the factors that determine the chances of life on other worlds? Are we special?

Then I want to look at some broader issues. Whether or not there is indeed life on other worlds, for example, could we ever travel there to find out for sure? If there are plenty of civilisations out there in the cosmos, have they ever visited us? Indeed, could they have influenced our development in ancient times? And finally, there is (for its own complex reasons) the most fashionable question of all as we approach the end of the century: are aliens visiting us now? And, if they are, why, how and what effect upon us would such a revelation produce? How would we feel if one morning we awoke to discover that the conspiracy theorists had been right all along?

In 1996 I wrote a book called *The Science of The X-Files*, in which I addressed a dozen different aspects of the paranormal – the stuff of the globally successful TV series, *The X-Files*. A couple of the topics in that book are tackled in greater depth in *Life Out There*, which will I hope please my existing readers and all those with genuinely open minds. However, it will further antagonise certain of those who reviewed *The Science of The X-Files* and those among my colleagues in the area of popular science writing who felt that I was in some way demeaning my training as a scientist or 'letting the side down', delving as I was into attempts to explain some aspects of the 'supernatural' using scientific means. This attitude demonstrates a depressing closed-mindedness that has become endemic within the scientific establishment. It is now *de rigueur* to be not merely impartial, willing to appraise the evidence with objectivity, but to expunge all attempts to clarify or broaden popular thinking about the so-called paranormal.

This was exemplified in a comment made by a hugely important and influential scientist called Enrico Fermi, who was one of the key figures in the development of the hydrogen bomb. His comment, made in 1943 to a group of friends who were trying to convince him of the high probability of extraterrestrial intelligence, was that if the Universe really was teeming with life, where was everybody?

Recently, this throw-away remark has become rather grandiloquently known as 'Fermi's principle', or sometimes 'Fermi's hypothesis', when it is merely a display of stupendous arrogance and ignorance little different from the pompous declarations of medieval popes.

Had Fermi not considered the position of humankind in the scheme of things? Was he not aware of the 500-year-long downsizing of the human condition, which had begun with Copernicus enlightening an egotistical race that the Earth was not in fact the centre of the Universe and ending with the random nature of quantum mechanics? Had Fermi chosen to ignore Darwin's theory of evolution via natural selection, demonstrating the insignificance of the individual and the erosion of the concept of a caring, controlling God? Worse still, did Fermi not consider how old our Universe really is, how vast the intergalactic void, how totally insignificant this lump of rock we call home? He should have done.

I hope that this book will offer a balanced viewpoint and present the facts, leaving you, the reader, to formulate your own interpretations. There is so much nonsense written about life on other planets, so much left out of the argument, so much one-sided presentation of the material (from enthusiasts and sceptics alike) that I thought such an attempt was overdue.

I hope that in my efforts to embrace all the issues surrounding the theme of life out there I have not glossed over the subject and that within these pages there will be something for everybody. It is a tale that perhaps uniquely encompasses a consideration of human psychology, special and general relativity, quantum mechanics, genetics, cosmology, space engineering and Darwinian evolution. If I achieve nothing more with this attempt, I will be delighted if my effort helps to keep alive the debate, encourages readers to support the on-going adventure, the great search, the dream and the expensive effort to answer a question that unites all humanity. After all, we live in a world in which all too many things divide us. And if indeed we are alone, perhaps the search for extraterrestrial life will encourage us to look after this place . . . and each other.

Gloucestershire
January 1998

1

LIFE ON MARS

'Under Mars is borne theves and robbers . . . nyght
walkers and quarell pykers, bosters, mockers and
skoffers, and these men of Mars causeth warre, and
murther, and batayle. They wyll be gladly smythes or
workers of yron . . . lyers, great swerers'

THE COMPOST OF PTHOLOMEUS

It was on a clear, sub-zero summer day in 1984 that geologist
Roberta Score made what may soon prove to be the first human
contact with an extraterrestrial life-form. She had finished the
day's work and was out with a group of colleagues, messing
around on snowmobiles in the Allan Hills region close to the
Antarctic base the team shared, when she saw a piece of black rock
in the ice and pulled up beside it to take a closer look.

This was Martian rock sample ALH84001 (named after the
Allan Hills in which it was found) – a piece of meteorite which,
twelve years later in August 1996, was to appear on the front
page of almost every newspaper on the planet and be proclaimed
the first example of fossilised biological material from another
world.

The project with which Dr Score was involved was an expedi-
tion financed by the American National Science Foundation's
Antarctic Meteorite Program, which had the specific task of find-
ing meteorites that had landed in Antarctica. Perhaps surprisingly,
there are huge numbers of meteorites in Antarctica, many of
which originate from deep space, and it has been estimated that

something in the region of one hundred tonnes of extraterrestrial material from Mars alone lands on the ice sheets each year.

Not knowing on sight what the sample was, Dr Score followed the usual routine for such finds – she bagged the meteorite and had it transported to the Johnson Space Centre (JSC) Meteorite Processing Laboratory in Houston, Texas, where it sat for eight years in a nitrogen-filled laboratory cabinet which had originally been built to house samples of moon rock brought back by the Apollo astronauts.

In 1993 a small slice of the meteorite was examined, and gases in the material were found to match those of the Martian atmosphere (a composition known about since the Viking missions of 1976). It was then neglected for a further year, until a team led by David S. McKay, working for NASA at the Johnson Space Centre with a special brief to search for fossilised biological material in Martian meteorites, removed another small fragment and set to work on it.

One of the first people to whom the team sent a minute sample of the rock was a British chemist named Simon Clemett, who was then working on his Ph.D. at Stanford University in California. The rock, sealed in a special air-tight sample bottle, arrived at his laboratory by FedEx in an inauspicious-looking brown envelope. In the accompanying letter, Clemett's contact at JSC asked him to analyse the sample and send back his findings, but they pointedly declined to say where the sample came from.

Treating the project as part of his Ph.D. work and just another job, Dr Clemett did the analysis, sent off his findings and forgot all about it.

It was two years later, and after he had returned to work in England, that Simon Clemett woke up one morning in early August 1996 to find his former supervisor from Stanford on the other end of the telephone advising him to get on the first plane to Washington and suggesting he take a look at the front page of the first newspaper he could find.

Clemett left his rooms through a gaggle of journalists on the doorstep. When he failed to stop to answer their questions, he was chased to the airport. Then, when the plane to Washington broke down on the runway and everyone had to disembark, he was again cornered by photographers and newsmen in the airport lounge. A few hours later, he finally made it to a hastily convened press conference, where he walked into a room filled with reporters from around the globe and a wall of popping flashes – moments before President Bill Clinton took the podium to announce to an excited world the findings of the NASA team.

News of Martian rock ALH84001 made the news-stands because of a press leak. The first suggestion that something extraordinary had been discovered in the Antarctic wastes was announced in a single-column news report that had appeared in the specialist magazine *Space News*. Noticed by a couple of vigilant reporters on a national paper, news of the NASA team's findings spread like wildfire and caught the researchers wrong-footed. They were planning to make the announcement in the periodical *Science* on August 16, but because of the amazing surge of interest in the implications of their discovery they were forced to hold a press conference immediately and to inform the President of their work.

At the Washington conference, Clinton announced that a special summit would be convened in November to discuss a review of the US space programme, and hailed the findings as a great discovery, adding that representatives at the planned conference would discuss how America should 'pursue answers to scientific questions raised by rock ALH84001. The rock speaks to us across all those billions of years and millions of miles. It speaks of the possibility of life.'

Following Clinton's announcement, representatives from NASA went into hyperbolic overdrive. Chief administrator Dan Goldin reinforced the positive buzz about the findings by saying: 'We are now on the doorstep to the heavens. What a time to be alive! In the last year we've discovered planets around nearby stars, we've probed to the depths of the Universe to see the formation and birth of galaxies. And today we are on the threshold of establishing if life is unique to planet Earth . . . We may see the first evidence that life might have existed beyond the confines of this small planet, the third rock from the Sun.'

But then, striking a note of caution, he added: 'I want everyone to understand that we are not talking about little green men. There is no evidence or suggestion that any higher life-form ever existed on Mars.'

Others supported the NASA claims with obvious delight. Donald Brownlee, an interplanetary dust specialist at the University of Washington, said: 'I think the team have a case that these things could be microfossils. It's unprecedented. It's one of the most important things in science, if it's true. Exobiology is intellectually interesting, but without any data it's just speculation. I think there's some data now.'[1]

Carl Sagan, for many decades a strong supporter of the search for extraterrestrial life and author of several books on the subject, was quoted as saying: 'If the results are verified, it is a turning point in human history, suggesting that life exists not just on two planets in one paltry solar system but throughout this magnificent Universe.'[2]

And at the headquarters of the Search for Extraterrestrial Intelligence (SETI), based in Mountain View, California, the organisation's president, veteran astronomer Frank Drake, gathered with his team around the television set, hooting, howling and cheering as the speakers took to the podium. Later Drake declared: 'It confirms what we've always believed – that life arises wherever the conditions are right. We are just one iota among countless iotas in the Universe.'[3]

NASA had fully expected a flood of responses from sceptics as much as the support of enthusiasts, and had even taken the precaution of calling in an eminent scientist, paleobiologist William Schopf, from the University of California, Los Angeles, who had maintained a sceptical view since he had first been informed of the research some time before the official announcement. Playing devil's advocate, Schopf said: 'I'm either an optimistic sceptic or a sceptical optimist. I do think this is a fine piece of work, and this is not easy science. This is multidisciplinary science. I happen to regard the claim of life on Mars, present or past, as an extraordinary claim. And I think it is right for us to require extraordinary evidence in support of the claim.'

Then, reviewing details of the structure of the fossils and the chemical and biochemical nature of the meteorite, he concluded: 'The biological explanation is unlikely.' And some others not employed by NASA concur. Monica Grady, curator of meteorites at London's Natural History Museum, and a member of a team who have studied Martian meteorites for over a decade, has commented: 'I think it's a very valuable piece of work. I just don't believe the final interpretation. They admit themselves that there are inorganic explanations for everything they have found.'

By this she is implying that the material on and in the meteorite which the NASA team claims to be biological in origin could in fact be nothing more than the product of non-living (or in this case, inorganic) systems. Yet, despite these arguments, the initial feeling about the discovery and its implications was one of surprise and excitement, with only a handful of dissenting voices.

The reason for the excitement is perfectly understandable. Ever since human beings had the luxury of leisure time in which to think and to wonder, we have looked at the stars and considered the possibility of life existing somewhere out there on other worlds. The theme of extraterrestrial life has been the mainstay of the great majority of science-fiction writing since the days of Jules Verne during the nineteenth century, and many science-fiction authors have focused their attention on Mars, the fourth planet from the Sun.

Mars was named after the Roman god of war, partly, it is thought, because of its colour in the night sky – the hue of blood. Later, in the seventeenth century, the astronomer Giovanni Cassini became the first to notice the white polar caps when he viewed the planet through an early telescope; and some century and a half later, during the 1830s, Wilhelm Beer and Johann von Madler noticed seasonal changes, the expansion and retraction of the polar caps and the changing colouration of the surface. Speculation grew that Mars might harbour plant life that showed seasonal variations, just as leaves on Earth darkened and fell and fresh buds grew – a 'wave of darkening', the two astronomers called it.

But the real excitement came in 1877, when Giovanni

Schiaparelli observed linear features on the surface of the red planet and called them *canali*. This word was wrongly translated from Italian into English as 'canal', when Schiaparelli actually meant 'channel'.

The American amateur astronomer Percival Lowell then took the idea of the Martian canal and created an elaborate image of a cold, arid and dying world on which intelligent Martians had constructed vast canals to carry water from one part of the planet to another to supply water for their farms. This, he claimed, accounted for the changes in colouration and the linear markings. He went on to declare: 'That Mars is inhabited by beings of some sort or other is as certain as it is uncertain what these beings may be.'

Sadly, this was nothing more than the work of an over-active imagination, and had absolutely no basis in fact. But Lowell's ideas did inspire an entire canon of science fiction dedicated to Mars, starting with H. G. Wells's *The War of the Worlds* (1898), in which malevolent Martians landed in England, followed by the influential but rather flaky novels of Edgar Rice Burroughs (the *Barsoom* series, begun in 1912), and culminating with Kim Stanley Robinson's modern classics, *Red Mars* (1992), *Green Mars* (1993), and *Blue Mars* (1996).

However, the dreams of the science-fiction writers are far from the rather prosaic truth. During the 1960s and 1970s both the Russians and the Americans sent a variety of increasingly sophisticated probes to Mars which beamed back images of an arid, freezing world possessing no hint of life. These missions began with Mariner 4 in 1964 (which flew close by the planet and found that the surface was pock-marked with moon-like craters), and ended with the Viking landers in 1976. These last visitors from Earth studied samples of the Martian soil and seemed to show that the planet was completely lifeless, its soil not only dead but apparently sterilised from the strong ultraviolet light streaming down on the planet's surface unprotected by an Earth-like atmosphere.

The data these missions provided may yet prove to present an overly bleak picture because, as we learn more about the extreme conditions in which life can survive, there is growing confidence in the idea that some form of very simple life could be found somewhere on Mars, perhaps deep beneath the planet's surface. Also, as we will see later, the evidence (provided by Vikings 1 and 2 in particular) is highly controversial, and the results of the experiments aboard these probes remain ambiguous.

Although the scientists working on ALH84001 hardly dare imagine the possibility that life may still exist in some hidden nook in the Martian polar regions or far underground, and would never suggest as much without conclusive evidence, there is the nagging but growing hope that such a thing may yet be discovered. Chandra Wickramasinghe, a physicist at the University of Wales, Cardiff, has said that such a possibility 'certainly can't be ruled out'. He went on: 'In fact I feel confident that [life on Mars] still survives. The lesson of microbiology over the past decade is that life can survive extremes that no one guessed before.' But a contradictory view comes from Ian Crawford of University College, London, who cites James Lovelock's 'Gaia' hypothesis: 'In Lovelock's models,' he says, 'life either flourishes after tailoring the environment to suit its needs or peters out. If life survived on Mars, it would have taken over the planet – it would not simply be hanging on.'

Mars is not the nearest planet to Earth – Venus is – but Mars has always been, and remains, the most likely place in our solar system other than Earth where life may have once flourished and may still survive. The surface of Venus is a scorching 500°C, and its atmosphere a cauldron of corrosive and toxic gases mainly comprised of carbon dioxide, which produces a dramatic greenhouse effect and is partly responsible for the high temperatures on the surface of the planet.

Mercury, with a mean distance from the Sun of around 35 million miles (about a third of the distance between the Earth and the Sun), is extremely unlikely to harbour any form of life. With a surface temperature only slightly lower than that of Venus, and its surface constantly bombarded by a destructive concoction of radiation from the close-by Sun, it would be difficult to imagine any form of life managing to get a foothold there.

Beyond Mars are the gas giants, Jupiter and Saturn, with atmospheres containing toxic gases constantly churned up by powerful magnetic fields. These planets offer little chance of harbouring life as we know it. The two largest satellites of the solar system, Titan, orbiting Saturn, and Ganymede, Jupiter's largest moon, might be more promising, and the Voyager probes that have passed close by recently have found what is believed to be organic molecules on the surface of Titan (see Chapter 3).

At the outer edge of the Solar System, Uranus, Neptune and Pluto would offer little comfort for what scientists refer to as carbon-based life-forms, because again the temperatures are either too low, their atmospheres noxious or, in the case of Uranus, the entire planet is a single ocean of superheated water, warmed by volcanic action and covered in poisonous gases.

At the time during which the rock sample ALH84001 may have been home to micro-organisms (between 3.6 and 4 billion years ago), conditions on Mars were very similar to those on Earth, and if we are to accept that life beyond Earth is a possibility, then this would be the time and the place when it could have appeared (just as it did on Earth).

So, given the historical perspective of at least a century of particular interest in Mars, and the fact that modern research has placed the odds of life on Mars as the highest in our neighbourhood (albeit still very low), there should be no surprise about the huge wave of excitement generated around the world at the news of the micro-organisms thought to have been fossilised in rock sample ALH84001. The simple fact is that human beings from all cultures and creeds find it almost impossible to accept the idea that we are alone in an infinite Universe. This in itself is such a chilling thought that to the agnostic it can be seen as one of the reasons mankind created the idea of God – so that we would have some company.

But, aside from the hopes and the wishes and the growing emotional investment placed on the findings of the researchers at JSC in Houston, what is the nature of the find, and how strong is the evidence for life having once found a home on the rock sample billions of years before finding terrestrial stardom?

There have been false scientific claims in the past. Extremists who question the very foundations of scientific investigation point to such diverse embarrassments as 'Piltdown Man' to illustrate what they see as the gullibility of scientists.* Another example was the so-called discovery of cold fusion in 1989, when scientists claimed they had created what was hoped would be an almost ever-lasting energy source – nuclear fusion in a test-tube.

Within the field of exobiology (the study of extraterrestrial life) there has already been at least one false claim to the discovery of extraterrestrial fossils. In 1961, a scientist named Bartholomew Nagy found what he thought were ancient Martian fossils on a piece of rock that had landed in France in 1864 – since dubbed the Nagy meteorite.

Nagy's investigation followed the same initial pattern as that conducted by the NASA scientists working on sample ALH84001, albeit with far cruder analytical techniques. But the claims were cast into doubt when it was found that the fossils were of nothing more extraordinary than terrestrial bacteria.

Within a week of the announcement of the findings of the team working on ALH84001, several groups of scientists and leading scientific commentators around the world were also beginning to voice doubts, and drawing what some believed to be unfair comparisons with the Nagy case.[4]

The discovery of micro-organisms on ALH84001 was quite different from those found on the Nagy rock in 1961. Ignoring the obvious differences in the detection and analysis equipment used in the 1960s and those of the 1990s, the material found on ALH84001 has far more impressive credentials.

ALH84001 is a potato-sized piece of meteorite weighing about 1.75 kg. It is thought that the rock, unremarkable in all aspects apart from the fossils it appears to contain, was formed at the same time as the rest of Mars, 4–4.5 billion years ago. This was the

---

* This was one of the most famous hoaxes in the history of science – the discovery in 1912 of 'Piltdown Man' (*Eoanthropus dawsoni*), the name given to a forged skull fragment deliberately manufactured to simulate the 'missing link' between apes and humans.

period when the inner solid planets coalesced and evolved into their present condensed states – bodies composed primarily of iron and other minerals (such as quartz and other silicon-based compounds).

Then, sometime between 3.6 and 4 billion years ago, liquid water seeped into fissures in the rock. This water, which is thought to have been in abundant supply on Mars at that time, was saturated with carbon dioxide from the Martian atmosphere. The water was therefore able to leave in the rock deposits of compounds called carbonates.

It is also believed that micro-organisms (which may have been living in the water) contributed to the production of these carbonates. These micro-organisms later died and became fossilised in the rock in much the same way that a variety of organisms from a wide range of geological eras have been preserved as fossils in limestone deposits here on Earth.

The rock probably stayed where it was, perhaps as a part of a cliff or a much larger outcrop, for billions of years. Then, around 16 million years ago, a giant comet or asteroid collided with Mars and our rock was torn away from the surface of the planet and sent hurtling into space.

ALH84001 then floated around in interplanetary space for almost all of the following 16 million years, while on Earth a seemingly insignificant variety of mammal evolved slowly into the ape family which, during the course of the next several thousand millennia, separated into a collection of species, one of which became *Homo sapiens*.

Much later, as early modern human communities developed agriculture, some 7–8,000 years before the construction of Stonehenge or the Pyramids, meteorite ALH84001 entered the Earth's sphere of gravitational influence and, around 13,000 years ago, crashed into the frozen wasteland of Antarctica.

There it remained untouched until Dr Roberta Score stumbled upon it during that fateful afternoon in 1984.

The fossils observed in ALH84001 are believed to be those of tiny, almost invisible primitive forms of bacteria-like organisms. The fossils are incredibly small: the largest of them is about

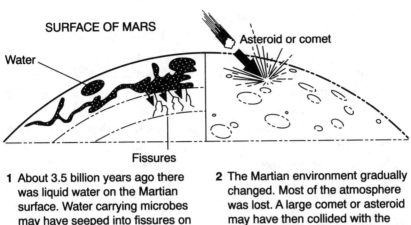

SURFACE OF MARS

Water

Asteroid or comet

Fissures

1 About 3.5 billion years ago there was liquid water on the Martian surface. Water carrying microbes may have seeped into fissures on the surface

2 The Martian environment gradually changed. Most of the atmosphere was lost. A large comet or asteroid may have then collided with the planet around 16 million years ago. This catastrophic event sent large pieces of rock into space

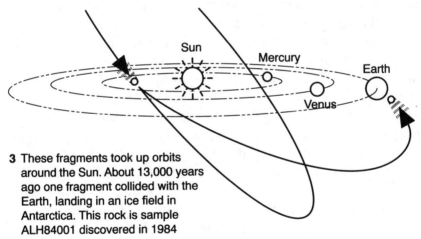

Sun

Mercury

Earth

Venus

3 These fragments took up orbits around the Sun. About 13,000 years ago one fragment collided with the Earth, landing in an ice field in Antarctica. This rock is sample ALH84001 discovered in 1984

Fig. 1: How ALH84001 arrived on the front page.

1/100th the width of a human hair, and most are ten times smaller still – so small that a thousand of them could be laid end to end across the width of the full stop which ends this sentence.

Some of the fossils are egg-shaped while others are tubular, but, strikingly, they are similar to fossils of bacteria and other micro-organisms found on Earth. In fact, they are so like terrestrial bacteria in appearance that when they were first seen that is exactly what they were mistaken for. The day they were discovered, one of the NASA team from JSC, Everett Gibson, took home the first pictures of the micro-organisms, put them on the kitchen table and his biologist wife immediately asked: 'What are these bacteria?'

So, what is the evidence to support the claim that these almost invisible 'dots' are in fact the first ETs encountered by humans?

The first positive aspect is that the rock sample definitely originated on Mars. Scientists are sure of this because the composition of gases found trapped in the structure of the meteorite matches that found in the Martian atmosphere as recorded by the Viking probes over twenty years ago. The main component of this mixture is carbon dioxide (CO_2), and as one of the British researchers who studied a sample of the rock commented soon after the announcement: '[The marked similarities] convinced people that [the samples] were Martian and we identified the major Martian atmospheric constituent, carbon dioxide. If you don't get carbon dioxide in high amounts, it is not from Mars.'

So, it is clear from which planet the samples originated, but what supports the idea that the biological material, as well as the geophysical structures, are of extraterrestrial origin? At first sight, the crucial piece of evidence is what appear to be the outlines of simple cell-like structures on the surface of the rock, mainly situated in deposits of compounds called carbonates, chemicals which are themselves carbon-based, often organic structures, but not necessarily biological in origin. However, the cell-like structures are extremely small, and so far no clear features such as cell walls have been visible using even the most powerful microscopes. These markings are so tiny they have been dubbed 'nanobacteria' (after the word *nano*, meaning objects in the size region of one thousand millionth, or 10^{-9}).

The third piece of evidence comes from deposits of inorganic compounds – crystals which contain iron, primarily iron oxide and iron sulphide. These are again produced by terrestrial bacteria. However, the deposits found in ALH84001 present a puzzle for the researchers. On Earth, some rare bacteria produce iron oxide in the form of magnetite, a compound which, as its name implies, is magnetic. It is thought that these organisms use them to help navigate – by aligning themselves with the natural magnetic field of the planet. But the fact that the Martian microbes appear to have also produced magnetite is a curious anomaly which has yet to be explained. This is because the magnetic field of Mars is no more than about 0.2 per cent (or 1/500) the strength of the Earth's magnetic field, so why would the micro-organisms on the rock produce magnetite?

The final piece of evidence, and perhaps the strongest of all in support of an extraterrestrial origin for the fossils, is the presence of large organic molecules never before seen in a Martian meteorite. These are called 'polycyclic aromatic hydrocarbons', or PAHs – oily compounds found in common materials such as coal and naphthalene (the major component of mothballs). PAHs are produced by decaying terrestrial bacteria, a fact which might initially suggest that these molecules found in ALH84001 are merely contaminants from Antarctica and not from Mars at all. However, the analysis of the sample shows a higher concentration of PAHs inside the rock than on its surface, which has lead Professor Richard Zare, head of the chemistry team that analysed ALH-84001, to say of this finding: 'Whatever else, these are the first organic molecules that have been associated with Mars.'

Yet others remain far from convinced and point to a series of anomalies in the results. First, it has been pointed out that PAHs have often been found in meteorites that have never been candidates for life (fossilised or otherwise), including rocks that have broken away from lifeless asteroids and fallen to Earth at one time or another. Geophysicist Robert Clayton has commented: 'PAHs are very widespread compounds in asteroids, and not diagnostic of life.' He goes on to comment that the concentration of PAHs in the Martian sample is around a thousandfold less diverse than those found in terrestrial fossils.[5]

However, in one respect this last comment could be seen to add weight to the argument for the micro-organisms having an extraterrestrial origin. It is almost impossible to say how differently Martian micro-organisms might have formed compared to their terrestrial cousins. Also, the fact that the range of molecules in ALH84001 might be due merely to a very different environment in which the rock formed and nurtured life actually argues *against* the idea that the rock had been contaminated. To some extent the narrow range of PAHs is perhaps indicative of the limited range of life found on Mars at the time when these micro-organisms could have formed (an argument that cannot be levelled against Earth).

McKay and his team have done their best to eliminate any chance of contamination creating the illusion of Martian life in the sample. They have done this by first ensuring that nothing was alive in the rock at the time it was first analysed, and secondly by demonstrating that the rock definitely originated on Mars. But the fact that there are more PAHs on the inside of the rock than the outside does not convince some of the critics, who point out that a black object such as the meteorite sample would absorb heat and melt snow around it, which would then seep into fissures in the rock and deposit terrestrial PAHs there. Furthermore, because PAHs are destroyed by ultraviolet light, those on the surface would have been more readily affected by the bright Antarctic sunshine. One American scientist has even gone so far as to say that McKay's arguments supporting his claims are 'flaky and simplistic. Weathering is a sloppy process. Things leach in, then leach out; they do not do the obvious.'[6]

Then, in January 1998, a highly publicised claim by a team led by Jeff Bada of Scripps Institute of Oceanography in California added what seemed to be a fatal blow to the findings of the NASA team. Bada's research indicated that the organic material found in ALH84001 had originated on Earth. '[This] work,' he declared, 'shows that the bulk of organic carbon in the meteorite is terrestrial and I say the compounds are terrestrial. It looks like everything is contamination.' However, even this statement was tempered with a controversial caveat: 'What we have shown is that there is no evidence in our hands that the meteorite contains any compounds that we could definitely trace to Mars . . . except maybe some tiny mysterious component that we don't understand at this point.'

Although the Scripps team were first to reach the press and therefore grabbed the headlines, their results are vehemently disputed by another group, this time from Britain. Led by Professor Colin Pillinger of the Open University, the team believe Bada is 'totally wrong', and that 'the evidence of the four samples of a second meteorite provides the most robust and uncontaminated evidence of organic material on the red planet. These cannot be dismissed.'[7]

Even so, there are other criticisms of the work of David McKay and his team, the most serious of which is the problem of the size of the micro-organisms.

Biologists believe that there is a lower limit to the size of a biological entity. This is simply because space is needed to store genetic material to allow the organism to grow, move and reproduce. On Earth, most bacteria are between 0.5 and 20 micrometres long (a micrometre is one millionth of a metre, or 10^{-6} metres). The objects discovered in the Martian sample have a size range of 20–100 nanometres (a nanometre being one billionth of a metre). In other words, even the largest of the objects is several hundred times smaller than the smallest conventional bacteria found on Earth.

Or at least so it has long been believed: researchers have recently turned up micro-organisms which are of a comparable size to the Martian objects. Microbiologist Todd Stevens, working at the US's Pacific Northwest Laboratory in Richmond, Washington, claims to have found bacteria just twice the size of the largest Martian examples. They are very odd organisms which live in rock fissures deep beneath the Columbia River in Washington. They have no organic food source and have no exposure to sunlight; instead they produce energy by a chemical reaction between rock and water, which creates hydrogen gas which they then use to convert carbon dioxide into methane – a chemical process with a net energy gain. Interestingly, all of these raw materials are known to have been present on Mars during the period in which the micro-organisms are believed to have been formed nearly 4 billion years ago.

Another researcher, Robert Folk, has also found bacteria fossils

similar to the Martian fossils, in hot springs in Italy. These he esti-
mates to be around 2 billion years old but about the same size as
some of the larger examples found in ALH84001.

A further objection concerns the nature of the carbonates in
which the fossils are found to be embedded in the sample. McKay
and his team believe these compounds were produced by biolog-
ical activity – waste products from the Martian organisms – but
others have described a mechanism by which similar compounds
could have been produced at the moment the rock broke away
from the surface of Mars and was hurled into space.

There is even an argument brewing over the age of the mete-
orite and the micro-organisms. The NASA team claim that the
microbes date from 3.6 to 4 billion years ago, but, based upon
experiments conducted by Dr Meenakshi Wadhwa of the Field
Museum in Chicago, the carbonates on the rock and the organisms
themselves might be only 1.39 billion years old, plus or minus 100
million years.

If this is shown to be true it would put an entirely different
complexion on the geochemistry and the biology of the sample
and consequently the micro-organisms found on it. But it does
nothing to disprove the fact that the markings might indeed be
fossils of a very simple ancient Martian life-form. We know very
little about how life could begin and evolve on other worlds, and
based upon the work of Robert Folk, there appear to have been
similar, apparently isolated organisms in existence on Earth
between the two extreme dates offered here (an estimate of 2 bil-
lion years for his organisms compared to NASA's 3.6–4 billion
and Wadhwa's 1.39 billion).

To rebut the arguments of the sceptics, McKay and his team
are currently conducting a further round of experiments. These
tests, some of the team believe, should have been conducted
before the first set of findings was made public. The head of the
chemistry group analysing the samples, Professor Zare, wants to
prove that the PAHs found in the rock are indeed from Mars and
not from Earthly contaminants. This seems the best way to qui-
eten the sceptics, but it is also incredibly difficult to achieve.

A further target for the team is to find amino acids or perhaps an internal structure to the cells they claim to have found on the rock. Amino acids are the building-blocks of life, the basic compounds found in all living things which form larger groupings called proteins and biochemicals such as DNA and RNA – the hallmarks of living matter (see the next chapter for an explanation of how these compounds inter-relate). The problem is that the cell-like structures McKay and the team have found are so incredibly small that searching for structures or components within them is presently beyond the limits of optical technology. The best hope for clarification seems to be to develop more powerful observational techniques to probe deeper into these tiny structures.

The other possibility is to back up the NASA claims with more fossilised microbes on other Martian meteorites. At the time of writing there has been only one other rather ambiguous find made by a British team, who, ironically, had been first to the microbes on ALH84001 but had been beaten to the announcement by the Americans.

When tiny sections of meteorite ALH84001 had been sent out to groups of researchers around the world in 1994, one such segment had arrived on the desk of Professor Colin Pillinger of the Open University (who in 1998 disputed the sceptics over the origin of the Antarctic meteorites). He worked on the analysis of the sample in collaboration with a husband-and-wife team, Dr Ian Wright and Dr Monica Grady, now of the Natural History Museum in London. The group had drawn a blank and found no evidence of organic materials in their sample. But that was by no means the full extent of their involvement with Martian meteorites.

Five years before the American team had begun to investigate ALH84001, Pillinger, Wright and Grady had been studying another similar meteorite, a rock named EETA79001, which had been found in Antarctica in 1979, five years before the discovery of ALH84001.

After intense analysis of EETA79001, the British team published their findings in the respected scientific journal *Nature*, and concluded that the rock contained high concentrations of organic materials, including molecules very similar to the carbonates later found surrounding the fossils in ALH84001. These were in particularly high concentrations deep within the rock; but, crucially, they decided that the risk of contamination by terrestrial matter meant they could not assume the rock contained evidence of fossilised Martian life.

A further hurdle for them was the fact that their rock sample was much younger than ALH84001. By their calculations, EETA79001 is only 180 million years old (about a twentieth of the age of ALH84001) and broke away from the surface of Mars as recently as 600,000 years ago (about the time *Homo erectus* was roaming the plains of Africa).

This entirely different time-frame meant that if they had indeed found evidence of Martian fossils, then life must have thrived on Mars between 600,000 and 180 million years ago, which to most researchers appears extremely unlikely given the hostile modern Martian environment. Because of these doubts, Pillinger and his co-workers limited themselves to a comment in their *Nature* paper that the findings had 'obvious implications'.

After the world awoke to the news of the NASA findings, Pillinger and his team were galvanised into staking their claim, and with some justification. Now, the NASA group have conceded that Pillinger and the others definitely made 'a very important contribution'. And one of those involved in the NASA research has said: 'The reason the more recent work on ALH84001 was done was largely on the encouragement of the earlier work the Brits and the others have done on Martian meteorites.'[8]

To be fair to the American investigators, the British findings are not as conclusive as those determined from ALH84001. But it does seem that on this occasion British reserve could have robbed the Open University and Natural History Museum researchers of what may prove to be one of the most important discoveries of the century, if not of all time.

So what are we to conclude from the claims and counter-claims, the evidence for and against ALH84001 containing genuine fossils? There seems to be only one clear consensus from commentators and researchers of all persuasions: before we can say with absolute certainty that the anomalies found on the Martian meteorite fragment discovered in the Allan Hills of Antarctica definitely indicate evidence of life originating from another world, a great deal more research has to be done.

Having said that, the weight of evidence is now beginning to shift further towards a terrestrial origin for the micro-organisms in ALH84001, although according to some dispassionate observers there remains an element of mystery surrounding the unusual nature of the fossils found in the rock.

Clearly, almost everyone who has heard about the discovery, not to mention those who have a vested interest in the outcome (for whatever reason), will wish that the markings, tubules and spheres in the rock fragments are the remains of something that once lived. And, of course, if irrefutable proof *is* found to confirm the terrestrial origin of ALH84001, this does not account for the findings of the British team working on EETA79001, and in no way refutes the suggestion that life may once have existed on Mars.

It is amusing to note that, since the discovery, the bookmakers William Hill have narrowed the odds they offer on the chances of finding intelligent life on Mars from 500–1 to 25–1. But, of course, wishing for something does not make it real, and William Schopf was spot-on when he said that extraordinary claims do indeed require extraordinary evidence.

2

WHAT IS LIFE?

'Who is to say we are not all Martians?'

RICHARD ZARE

What is life? At first glance, the answer might seem obvious, but it is in fact very difficult to produce an all-embracing, logically consistent definition of 'life'.

If we begin by suggesting that all living things grow and move, this does not help much. After all, crystals grow – they produce regular patterns, repeated simple units which might easily be compared to the physical appearance of cells. Inanimate water or any other liquid can flow, or move, so in itself this ability is also clearly insufficient to define life.

Perhaps a more sophisticated criterion is to say that all life-forms use energy – but then so do all machines, from a lawn-mower to computers, cars or interplanetary vehicles. A slightly more useful definition might be to say that all living beings have the ability to *control* energy – but then so do some advanced machines such as those developed in recent years that utilise the concept of fuzzy logic.

One intriguing example which illustrates the difficulty of defin-ing 'life' is to imagine what alien observers would make of the Internet, if they detected it before noticing the humans who use it. The cyberneticist Kevin Warwick describes the reaction of an alien race to the Internet by asking a series of questions:

How many of the seven tests of life does it actually pass?

Growth, certainly; indeed, the growth of such networks over the last few years has been tremendous. Movement, definitely; e.g. the movement of switches in the network. Irritability, certainly; it is the network's role to respond to stimulation. Nutrition, yes; messages (energy in one form or another) are entered into the network. Excretion, yes; messages are passed out of the network. Breathing, more difficult, but yes, if we consider electronic pulses passing around the network. Finally, reproduction, the most difficult of all to argue, might be inferred from the fact that an original network starts off a new network elsewhere.[1]

An alternative way of looking at the meaning of the term 'living' is to argue that only living things process and store information. But is this not the sole purpose of a computer? The debate about the possible future development of artificial intelligence by complex computers still rages (and will be dealt with in Chapter 3), but present-day computers could in no way be seen as living beings, yet they process information. So how can we pin down the quintessential factors that separate the living from the inanimate?

Using the old-fashioned school-book definition – that all living things exhibit the three 'f's: fight, flight and frolic – still leads us into logistical problems. Frolic is of course a euphemism for 'reproduce', but a flame 'reproduces', as do inorganic crystals growing in a solution. Probably the best we can do is to say that all living things, from the simplest bacterium to a human being, reproduce and pass on genetic material or inherited characteristics to their offspring, and that this material has undergone some degree of mutation. In other words, they have taken part in the evolutionary process via natural selection and haven't simply produced exact copies of themselves.*

---

* This naturally raises all sorts of ethical considerations, particularly in the light of recent genetic breakthroughs involving cloning. For instance: by this definition, is the cloned sheep, Dolly, a living thing? A clone is after all an exact replica of its parents. Although fascinating, this matter is of such labyrinthine complexity that it deserves a book of its own, and is beyond the context of this single chapter.

Shortly before his untimely death, the late Carl Sagan defined life as 'any system capable of reproduction, mutation and the reproduction of its mutations'. What he meant by this was that life was represented by any entity that allows for variation from generation to generation using the mechanism of evolution via natural selection, an entity that can pass on its characteristics, reshuffled by the processes of reproduction, so that those characteristics will not appear to be exactly the same in the next generation.

We will return to a detailed discussion of evolution in the next chapter, but for the moment this definition of 'life' will have to do. (It is perhaps more a question of semantics rather than practical science, anyway.) What is surely of far greater importance is the question: how did life originate on the one planet upon which we know it flourishes in great abundance – the Earth?

In order to approach this matter we need to look at some basic concepts behind the broader picture of how life could have arisen on the early Earth, and how this process could have occurred elsewhere in the Universe and at different times in its history.

All matter consists of atoms. There are over one hundred different atoms, some of which are common and everyday, such as oxygen, nitrogen, iron and lead; others are less so – substances with strange names such as rubidium, einsteinium and selenium. The most important for any discussion of life is the carbon atom. In many respects carbon is like any other atom: it is stable, it forms bonds with different atoms and with other atoms of carbon to form molecules – conglomerates of atoms which vary in size between those containing a few components to molecules consisting of millions of atoms. But in one vital respect it is different from any other atom. It has a property which makes it the only known atom which can form the backbone of really large molecules (organic molecules) and even larger conglomerates (biochemicals). The late writer and chemist Primo Levi illustrated the versatility of carbon and its intimate involvement with life in a famous piece he wrote about the element:

Our character lies for hundreds of millions of years, bound

to three atoms of oxygen and one of calcium, in the form of limestone . . . A blow of the pickax detached it and sent it on its way to the lime kiln, plunging it into the world of things that change . . . It was caught by the wind, flung down on the earth, lifted ten kilometres high. It was breathed in by a falcon . . . dissolved three times in the water of the sea . . . and again was expelled . . . then it stumbled into capture and the organic adventure . . . It had the good fortune to brush against a leaf, penetrate it, and be nailed there by a ray of the sun . . . In an instant, like an insect caught by a spider, [the carbon atom] is separated from its oxygen, combined with hydrogen and finally inserted in a chain of life . . . [It] enters the bloodstream: it migrates, knocks at the door of a nerve cell, enters, and supplants the carbon which was part of it. This cell belongs to a brain, and it is my brain . . . and the cell in question, and within it the atom in question, is in charge of my writing, in a gigantic minuscule game which nobody has yet described. It . . . guides this hand of mine to impress on this paper this dot, here, this one.[2]

Carbon has an almost unique ability to form long chains and rings of atoms around which other atoms can be attached. I say 'almost' because there are other atoms which can form chains and rings, but these do not exhibit anything like the variety or the range of carbon-based molecules. Silicon, an atom which shares some of the characteristics of carbon, is the closest example, but because the bonds formed between silicon atoms are not so strong as those between carbon atoms, it can only form stable chains of up to five or six atoms in length.

Silicon also fails to match another incredibly useful property of carbon. Carbon has the ability to form multiple bonds with other carbon atoms and sometimes with other suitable atoms. This adds greatly to its versatility, allowing it to form a vast variety of molecules of different type, some of which are truly huge, consisting of millions of atoms. Silicon cannot form multiple bonds with its own type and does so only rarely with any other atom.

Adenosine triphosphate (APT),
a biochemical based on carbon

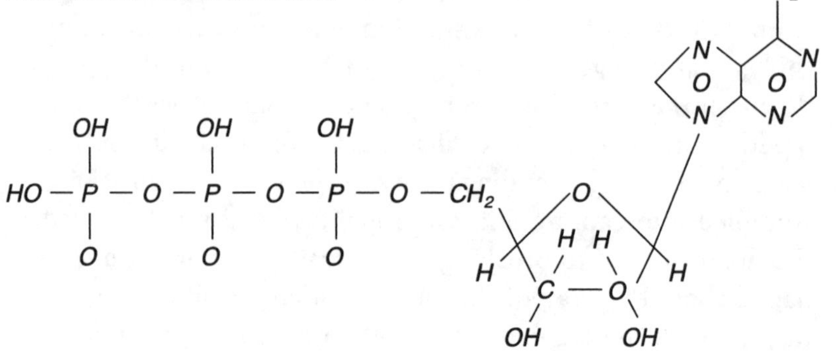

One of the most complex
molecules silicon can form
(in the way carbon forms
organic molecules) – far
less complex than those
possible using carbon

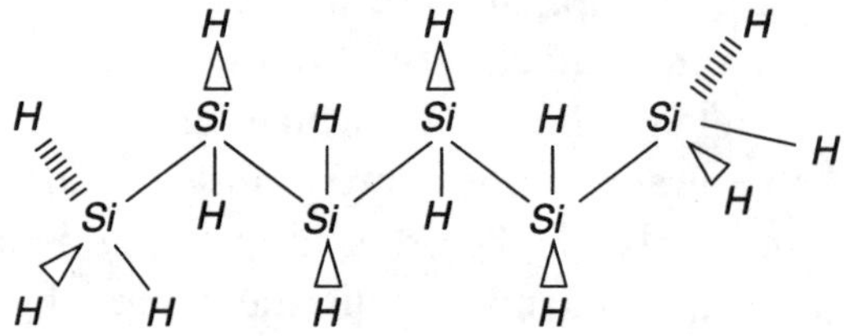

Fig. 2: Molecules based on carbon compared to those based on silicon.

So, despite the fact that some science-fiction writers have had a good stab at describing situations in which a silicon-based life-form could exist, this is extremely unlikely, simply because, according to our present knowledge of chemistry, silicon cannot form molecules anywhere near complex enough to sustain life. Carbon is unique in our Universe. It is the only atom able to form the building-blocks of life.

But, you may wonder, isn't this a little chauvinistic? How can I say that carbon is unique in this way when I have never travelled beyond the Earth, when the species I am a part of has never ventured beyond the Moon, and has hardly begun to explore the planets within our own tiny solar system using simple robot craft?

The answer to this is an important factor in understanding how science can arrive at answers concerning the matter of life on other planets. One of the most basic rules of science, 'the principle of universality', states that the Universe is homogeneous – or, to put it a different way, 'what happens here, happens there'. For example, we know that there cannot be another atom like carbon that we have inexplicably missed, for the simple reason that such an atom would not fit into what is called the periodic table – a scheme in which all the different types of atom in the Universe have a strict position and interconnect in a precise pattern.*

The periodic table was devised over a century ago by a Russian chemist named Dmitry Ivanovich Mendeleyev, who placed the elements into inter-linking patterns according to their character-istics – a grid system called 'groups' and 'periods'. These atoms vary in size according to their atomic number, and there are no 'spaces' into which some odd element – perhaps found only in the constellation of Orion – would fit.† Over the decades since it was first established, all the gaps in the periodic table have been filled, and scientists have *extended* the periodic table, but they could never discover any other previously unknown element that some-how fits into the middle of the scheme.

The point of this diversion is to establish that carbon is the only atom that can form molecules large enough to act as 'the mole-cules of life' – the enormous structures such as DNA (deoxy-ribonucleic acid) and RNA (ribonucleic acid), or even the smaller

---

* Some of you may still be unconvinced and suggest that perhaps we only think that the Universe is homogeneous. The argument against this is complex and could con-stitute an entire chapter in itself, but one example will, I hope, suffice. Using a technique called spectroscopic analysis, astronomers are able to detect light impulses from distant stars and analyse their constituents. This process illustrates that the light from stars is similar to that from our own Sun (varying principally in the pro-portions of different colours). This light is produced by changes in the chemical nature of the elements of which the stars are made, and can be matched with ele-ments found in our own Sun and indeed here on Earth, as Primo Levi conveyed in the earlier quote. At their root, all things – the bricks of buildings, the cells of our eyes and the paper of which this book is made – come from stars, and the funda-mental components of stars are universal.

† At the last count, there were 109 elements in the periodic table, but new additions are made from time to time as a result of investigating the by-products of nuclear reactions. These are called transuranic elements and are short-lived, unstable atoms that quickly decay to smaller elements. Most importantly, they are all heavy ele-ments which are added to the *end* of the table, atoms of higher atomic number and atomic weight than the stable elements which make up the bulk of the table.

building-blocks, the nucleotides that go to form these massive structures, and the proteins, enzymes and other biochemicals needed to run cells and sustain our existence. And, because of the homogeneous nature of the Universe, this is the case both 'here' and 'there'.

So what of these biochemicals? What is so special about them?

Biochemicals perform a staggering range of functions, from the provision of energy for the maintenance of every cell in the body of any living thing, to the essential prerequisite of 'life' – the processes of reproduction. There are thousands of these biochemicals, each specifically tailored by nature or evolution to the point where they can perform a particular task with amazing efficiency. But probably the most complex task performed by a particular group of large biochemicals is the elaborate series of steps involved in the passing-on of genetic information from one generation to the next.

As I said at the beginning of this chapter, the ability to reproduce, and in so doing, to pass on genetic information that also has the ability to mutate, is perhaps our clearest way of defining life. And this process relies upon the employment of large organic molecules or biochemicals. For this reason alone, it is very difficult to envisage any life-form, and in particular any life-form which has developed intelligence or civilisation, to be based upon any other framework than chemistry dictated by the role of carbon.*

So, having come to the conclusion that the ability to reproduce and to allow for mutation from generation to generation lies at the heart of what we mean by 'life', and that biochemicals play a key role in this mechanism, how exactly are they involved?

At the root of the subject is the simple fact that all life-forms pass on characteristics to the next generation via a genetic blue-

---

* To some, this may again sound chauvinistic, and this time such a comment would be more justified because there is the possibility that life-forms exist in the Universe which developed in some as-yet incomprehensible way, but such a possibility is for the moment beyond the scope of my argument. It will, however, be addressed in Chapter 4.

print stamped into every cell in every living creature. The blue-print is different for every individual. This is what distinguishes a monkey from a frog or Arnold Schwarzenegger from Tony Blair. The blueprint is called the genetic code, and is made up of a sequence of tiny units called genes which are themselves made from bits of a very large biochemical named DNA (deoxyribonu-cleic acid).

Genetics goes back a long way. It was an Austrian monk named Gregor Mendel who, during the mid-nineteenth century, realised that characteristics were inherited from generation to generation by means of what he called 'discrete factors' – what today we call genes. He realised that each individual inherited two complete sets of genes called 'alleles', one from each parent, and that these alleles are not changed during reproduction but are passed on unaltered from parent to offspring. What makes us all different is that, because each parent in turn has two alleles, each offspring has a 50 per cent chance of getting one or other allele from each parent, or a 25 per cent chance of receiving any particular combination of genes. This leads to a 'shuffling' of genes and produces variations in characteristics as diverse as colouring, build, suscep-tibility to certain diseases and even such traits as a tendency towards alcoholism.

Although Mendel worked out the idea of inheritance of genetic material, he had no idea how the process worked on a chemical level. It was not until almost seventy years after his death that the mystery was solved, when James Watson and Francis Crick, work-ing in Cambridge, discovered that genes consisted of two strands of a complex molecule, DNA – the now-famous double helix.

DNA is found in every cell of every living creature. And, although it is a very large organic molecule, amazingly, it is all made from just four small chemical units or bases, called A (for adenine), T (thymine), G (guanine) and C (cytosine). There are hundreds of millions of these four bases scattered throughout the structure of each molecule of DNA. They combine three at a time to form a specific code or 'word', and these three-letter codes allow individual amino-acids to be positioned in an exact sequence to form proteins. Proteins not only make up the structure of humans but are also responsible for all the complex chemistry needed for what we call life.

The best way to visualise how this works is to imagine a cell as being a huge set of encyclopedias. Each individual volume in the set is equivalent to a chromosome within the cell. Each human cell has twenty-three pairs of chromosomes, made up of an incredibly long, tightly coiled length of DNA. So the human 'encyclopedia set' would consist of forty-six volumes. Each volume in this library would be billions of words long.

Just as each volume of the encyclopedia deals with a plethora of subjects, so each chromosome controls every physical characteristic of the organism. Thus, in this analogy, eye colour might be equivalent to, say, 'Renaissance Economic Theory'; hair type to 'London Double-decker Buses in Use During WWI'; and height to 'The Mating Habits of the Duck-billed Platypus'. These individual entries in the encyclopedia are equivalent to individual genes. And, of course, encyclopedia entries are made up of paragraphs, words and letters. So, extending our analogy, the paragraphs are

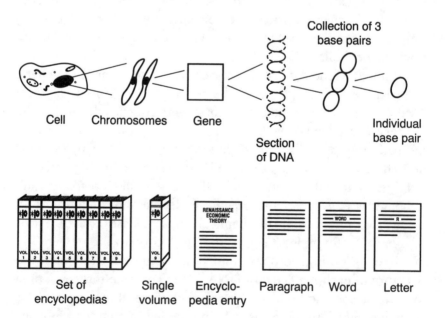

Fig. 3: The analogy of the genetic code to words, books, and libraries.

equivalent to the large sections of DNA which make up specific parts of the genes. Words are the counterpart of the three-letter 'words' which code for the individual amino-acids, and the letters are the base pairs, A, T, C and G.

As well as providing the material from which genes are made, DNA also acts as the vehicle for growth by replicating itself within the cell. It acts as a template to produce new copies of itself. It is as if the encyclopedias in our collection could be end-lessly photocopied and distributed; any number of exact copies of each can be made by unbinding them and generating copy after copy based upon the original. But, as with photocopies, tiny mis-takes occur during the copying. These 'mistakes' are called mutations, and they can either be beneficial or detrimental to the offspring.

Within fifteen years of Crick and Watson's ground-breaking work on the structure of DNA, scientists had a pretty clear idea of how the words in the encyclopedia collection could be formed. They realised that out of the four bases, A, T, C and G, only three were needed at any one time to correctly position an amino-acid to form a protein. This meant that with a supply of four letters, there would be sixty-four ways in which three-letter com-binations could be created. For example, the combination 'G–A–T' specifies the amino-acid, aspartic acid.

However, even this is not the end of the story. DNA is not the only biochemical involved in the mechanism of genetics. A com-pound of equal importance is a close relative of DNA called RNA (ribonucleic acid). Now, in our encyclopedia and photocopy analogy, the pages of the encyclopedia would be reproduced by a mechanism in the photocopying machine – it scans the page and translates that to another page upon which the ink is drawn and the repeat image made. The biological version of the device that carries the information from the master copy to the copying ele-ment itself is the RNA molecule. This very special molecule may be thought of as a messenger – it carries the code from DNA to a biological 'device' in each cell called a 'ribosome' (equivalent in our analogy to the actual duplicator element of the photocopier), which constructs proteins.

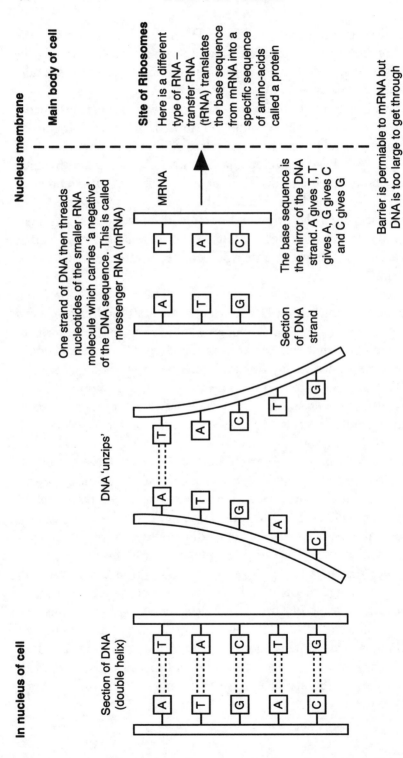

Fig. 4: How DNA produces proteins.

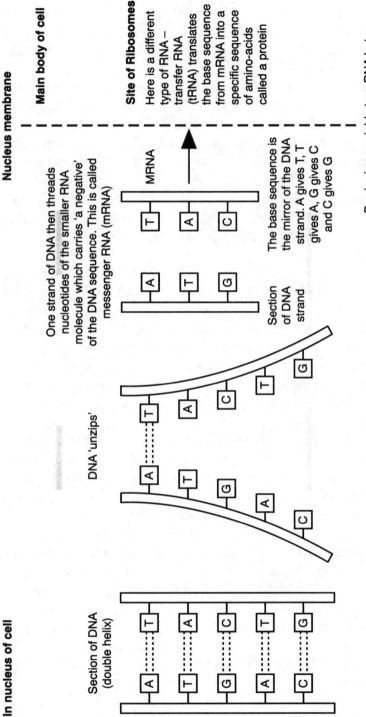

Fig. 4: How DNA produces proteins.

The entire process can be visualised as a little like a production line in a factory. DNA makes more DNA and RNA from what are called 'primed nucleotides' – large organic molecules which have been pre-prepared by the body to interlock with each other in the correct way. The analogy would be that these primed nucleotides are like bits of machinery made in a machine shop which can be fitted together on an assembly line. To perform this task, DNA also uses enzymes, which are other types of organic molecule (proteins) which facilitate certain specific reactions (they speed up the reaction; in other words, they act as biocatalysts). The RNA then transplants the design instructions from DNA into the ribosomes which make proteins using a supply of simpler components called amino-acids. The proteins then form more DNA, RNA and other larger molecules, while others act as enzymes and take their part in the mechanisms that maintain the cell and therefore the body of the living being (including the above process).

This then is a self-sustaining production line. The brain is the chairman of the company, the nucleus of every cell in the body is the shop floor, the large organic molecules are equivalent to the machines, made from the same parts that are being manufactured, all powered by the food we eat. The shop floor is ventilated by the air we breath and all the processes take place in water (or an aqueous solution).*

This all appears to be very neat. It explains the mechanisms through which life sustains itself. But isn't there something wrong with this biochemical description of the cycle of life? Consider the elements of the mechanism: 'life' is defined as a system which can reproduce and pass on mutated information from generation to generation or an entity which is involved in evolutionary change via natural selection. Evolution is facilitated by genetic variation through reproduction. The genetic code is carried by DNA, which is made in a biochemical process in the cell in which proteins are formed using RNA. But, and this is the crux

* This last factor is an important reason why water is of paramount importance in almost any environment in which living things can maintain a foothold (see Chapter 3).

of the dilemma, if the ability to partake in evolution is a requirement of 'life', and this process itself requires an elaborate set of processes, how did 'life' originate in the first place? To put it another way: any entity that can evolve (or, by our definition, be 'alive') has to be complex enough to possess the genetic machinery via which it can evolve – even the simplest bacteria possess this (it is why bacteria are consider to be living). But how could an entity became this complex without evolution? How could it get on to the evolutionary ladder at all? It is a classic 'chicken and egg' dilemma. But, luckily, this seemingly impossible conundrum does have a solution.

Until the seventeenth century, all ideas about how life could originate were firmly rooted in religion rather than natural philosophy (the precursor of science). God was seen as the sole progenitor and the controller of all life; humankind was seen to have been created by divine processes which, it was presumed, could never (and perhaps should never) be understood by mere mortals. But then gradually theories began to emerge that attempted to explain how the trick could be performed.

One of the earliest ideas was the concept of 'spontaneous generation'. Since ancient times it had been observed that some organic materials, particularly foodstuffs such as cheese or bread, appeared to grow organisms on their surfaces without any noticeable means. For example, a piece of cheese left out for a few days grew a mould, as did bread or over-ripe fruit. This led to the idea that life could arise 'spontaneously'.

It was only in 1860 that the idea lost its impetus when Louis Pasteur showed that if these materials were placed in an air-tight vessel they did not grow any form of fungus or visible film. But even then, supporters claimed that all Pasteur had shown was that any life spontaneously arising on the cheese or the bread was being suffocated or prevented from growing because of lack of air. Pasteur then demonstrated that in fact the fungus or the bacterial growth was produced by invisible spores or micro-organisms present *in* the air – the appearance of a life-form on the surface of the organic material was due to the colony of bacteria or the fungus growing to the point where it was visible with the naked eye, and that air, moisture and a reliable food supply were all necessary for this to happen.

The death-knell for the theory of spontaneous generation came almost contemporaneously with Pasteur's work, when Charles Darwin and Alfred Russel Wallace simultaneously developed the theory of evolution via natural selection. In his revolutionary work, *On the Origin of Species*, Darwin showed that life evolved from simple to more complex forms over long periods and that life, however simple, could not spontaneously appear on a piece of cheese. Bowing to the religious sensibilities of his time, Darwin did not in *On the Origin of Species* go so far as to speculate that life had evolved on Earth without some form of divine intervention. In the book he said that 'the Creator' had breathed life 'into a few forms or into one'. However, in private correspondence with his friend and fellow scientist Joseph Hooker, Darwin expressed what he really thought – that life had arisen from chemistry: 'In some warm little pond with all sorts of ammonia and phosphoric salts, light, heat, electricity, etc. present.'

The idea of extending the concept of spontaneous generation to the origin of life on Earth, or any other planet, has as much validity as the early experiments of natural philosophers. All life evolves, and all life as we can conceptualise it must, by definition, operate within the law of natural selection – it has to be wrapped in the process of genetic mutation over successive generations and has to be complex enough to become involved in this process. Although we appear to have a chicken-and-egg problem, the route of spontaneous generation is no answer, just a blind alley.

One way of side-stepping the origin issue is the idea of 'panspermia'. This concept first became popular in the nineteenth century due largely to the Swedish chemist Svante August Arrhenius, who suggested that life reached Earth by means of spores from outer space. This idea was supported by many other respected scientists of the time, including the eminent physicists Hermann von Helmholtz and Lord Kelvin. In its most elaborate form, panspermia proposes that 'the seeds of life' pass from planet to planet and that soon after a planet evolves to the point where it can support life, it will.

According to the adherents of the theory, life developed on
Earth some 3.5–4 billion years ago, after seeds or spores arrived
here in the form of simple organisms. At that point in the history
of the Earth, the planet had evolved a suitable atmosphere and a
conducive environment which allowed these 'seeds' to flourish and
to embark along the path of Darwinian evolution.

This theory remains highly controversial. The most vocal sup-
porters of the modern version of panspermia, Fred Hoyle and
Chandra Wickramasinghe, have been arguing their case for the
best part of three decades, and indeed, there is no hard scientific
reason why this theory should not be true. We know that the
cosmos is filled with organic material. Even if it is the first exam-
ple of an extraterrestrial rock containing fossils of life-forms, the
Martian sample ALH84001 is certainly not the only meteorite
ever found to possess organic material of one form or another. A
great variety of often complex organic molecules have been found
in rocks which have fallen to Earth, some of which are thought to
have been in space for tens of millions of years, rocks which may
have originated from locations far beyond our solar system. The
same is true for comets which are known to travel vast distances
and which (in admittedly rare cases) could have originated in the
solar systems of other stars.*

One argument proposed by those sceptical of the concept of
panspermia is that bacteria, which are seen as the simplest form of
life, could not survive a journey across the cosmos lasting perhaps
many millions of years. Yet new evidence is taking the shine off
their protests.

During recent years bacteria have been found in some surpris-
ingly hostile environments on Earth. A decade ago scientists were
shocked to discover bacterial growths close to active volcanoes.
Similarly hardy organisms have since been discovered living at the

---

* Most comets observed so far are thought to come from at least two distinct regions
some distance beyond the edge of our solar system, called the Oort cloud and the
Kuiper cloud. It is not inconceivable that other celestial objects that have made con-
tact with the Earth regularly during the past 4.65 billion years have originated on
other worlds orbiting distant stars.

extremes of temperature on Earth – in the Antarctic and Arctic wastes and in hot springs. Further varieties have even been found flourishing in the vents of nuclear reactors. But a discovery made in 1996 puts all these earlier findings in the shade. Scientists drilling into the seabed off the eastern coast of the United States found bacteria living 2,500 feet beneath the sediment, which itself lay under 11,000 feet of water.

It is difficult to exaggerate just how astonishing this discovery is. The environmental conditions over 13,000 feet below sea-level are far from those experienced by most living beings on the surface of our world. The temperature down there would be around 170°C (380°F), and the pressure approximately four hundred times greater than it is on the surface of the planet. Although different from the conditions found in interstellar space, where the temperature verges on absolute zero (– 273°C or 0K) and there is naturally no atmospheric pressure, these conditions are certainly as harsh for living things. It would not be unreasonable to suggest that any bacteria hardy enough to survive 13,000 feet below the Earth's surface would be able to tolerate the conditions found in space, if it was held in some form of suspended animation in the heart of a comet or meteor.

However, such contortions of what may or may not be possible might be quite unnecessary. Earlier in this chapter I said that DNA and RNA are produced by what are called 'primed nucleotides', large organic molecules or biochemicals which slot together to form more substantial units that produce the components of DNA and RNA. Some argue that seeding could have occurred via these primed nucleotides, that these simpler, 'prebiotic' units (large organic molecules that act as precursors of living organisms but which are not themselves 'alive') could have made the journey across space to seed life on Earth.

This 'kick start' could have come from elsewhere in our solar system – perhaps Mars, if life did indeed flourish there immediately before it emerged on Earth. Alternatively, the progenitors could have come from a far distant sun many light-years away, where life had originated earlier than it did within our solar system.

Although the distances between the stars are vast, and the journey time for organic material locked inside a comet or meteor would be at least millions of years, it is not inconceivable that life could have been seeded here in this way. The latest estimates place the age of the Universe at around 12 billion years, and the earliest recorded life on Earth at between 3.5 and 4 billion years ago. Which means that there was a gap of at least 8 billion years between the Big Bang and the first appearance of life on Earth. It is quite likely that other planetary systems elsewhere in this galaxy, or in others, began to form and cool slightly earlier than ours and, if life began to evolve there just 1 per cent earlier than it did here, this allows for tens of millions of years for that seed to reach our planet.

A further refinement of the panspermia idea was proposed in the 1970s when a group of Canadian and British scientists developed the 'life cloud' theory, which describes how organic chemicals known to exist in stellar dust clouds could react to form nucleotides and proteins. According to this theory, when planets pass through these clouds, the seeds of organic life are planted, and if the planet is in a suitable state to act as a cradle for life to flourish, then it will. The Earth, it is suggested, may have passed through just such a cloud around 3.5 – 4 billion years ago.

The possibilities offered by these various permutations of the panspermia theory are intriguing, and the idea that life on Earth was seeded on some far-off world does not raise insurmountable scientific objections, but neither does it offer a true solution to the mystery of how life may have originated. All these ideas do is move the goalposts. If we assume for a moment that the theory of panspermia is correct and does indeed account for how life began here, we are still left with the question: how did life begin on the first world on which it appeared?

Whether the theory of panspermia is one day proven to be correct or false, if we are to get anywhere in determining how life began, we have to put the concept aside. Instead, let us assume for the moment that life on our planet was not seeded and that it originated here by some mechanism linked to the environmental conditions of the time. How could it have happened?

One of the earliest attempts to describe a biochemical process in which increasingly complex molecules could be formed from a soup of simple chemicals appeared in 1936, in a book entitled *The Origin of Life* by a Russian biochemist named Aleksandr Ivanovich Oparin. In his book, Oparin proposed that primitive organisms could have arisen from pre-existing organic compounds that had themselves developed naturally from simpler compounds. In this way, he was really applying ideas of Darwinian evolution to pre-biotic systems. But despite being able to support his ideas with meticulous laboratory experiments, Oparin's theories were, perhaps not unexpectedly, greeted with almost universal scepticism from religious thinkers and scientists alike. However, he was supported by the great British biologist, John Haldane, who was the first to propose the idea that certain organic chemicals found in primitive oceans could have combined to create self-reproducing forms. This, Haldane suggested, would have been encouraged by an atmosphere rich in hydrogen – a set of conditions known as a 'reducing atmosphere'.*

The idea that life may have arisen upon a primitive Earth which had an atmosphere rich in hydrogen was fashionable throughout the 1940s and 1950s, and was given a tremendous boost by a now-legendary experiment conducted in 1953 by two American chemists, Stanley Lloyd Miller and Harold Clayton Urey.

In April 1953 the world was stunned to learn that Crick and Watson had elucidated the molecule structure of DNA – the 'double helix' mentioned earlier. Suddenly the subject of life and how science could begin to illuminate the mysteries of genetics and evolution were hot news, and it was into this environment that a month later, in May 1953, Miller and Urey published their revolutionary findings.

With their experimental arrangement, Miller and Urey were trying to verify the theories of Oparin and Haldane – an experiment set up in the laboratory which would attempt to duplicate

---

* Based upon the fact that hydrogen is a molecule which undergoes a chemical reaction called reduction, the opposite of oxidation.

what they believed were the environmental conditions on Earth at the time at which life originated. They combined a gaseous mixture of hydrogen, water, methane and ammonia (to match the components and proportions of the primordial atmosphere), and, to simulate lightning, they subjected the blend to an electric discharge for one week.

For the time, their results were truly staggering. In the bottom of the reaction flask Miller and Urey found a red-brown sludge that turned out to contain significant amounts of organic compounds. These included a variety of organic acids (aliphatic acids and hydroxyl acids), urea, and a set of compounds that had similar structures to carbohydrates. Further experiments showed that a wide range of molecules essential for life could be formed in this way.

Following this initial success, the pair added another simple molecule to the original gaseous mixture, hydrogen cyanide (HCN), found in volcanic gases, and it was then that they made

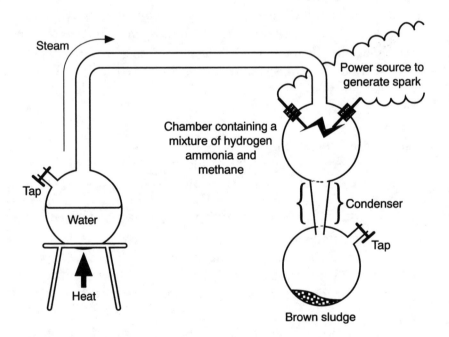

Fig. 5: The Miller–Urey experiment.

their most significant discovery. This time the sludge they produced contained a collection of one of the types of compound known to play a key role in biochemical mechanisms and almost certainly a key player in the origin of life – amino-acids.

There are twenty different types of amino-acid found in nature, and these combine in various ways to form proteins which can, among other things, act as enzymes and become involved in the processes that form the nucleotides that combine to produce DNA and RNA. In short, amino-acids are the fundamental building-blocks, molecules that lay at the very heart of almost all the processes in the cell. Miller and Urey found no fewer than eight of these twenty amino-acids in the reaction flask.

Their conclusion was that a 'soup' of life-forming molecules could have been brewed in the Earth's atmosphere during the space of just a few years, and recently, Stanley Miller has declared that these molecules could have developed in complexity and produced living cells within perhaps as little as 10,000 years. Flying in the face of critics who claim life on Earth is unique, and based on his own experiments, Miller is sure that given the correct environmental conditions and the proper blend of chemicals, life could form on any planet.

Miller and Urey's work was given further support eight years later when, in 1961, a biochemist named Juan Oró working at the University of Houston tried to see if amino-acids could be formed using even simpler chemistry. He used just two of the original Miller–Urey ingredients – hydrogen cyanide and ammonia in an aqueous solution. To his surprise, he not only formed a significant number of different amino-acids, but also an abundance of adenine, one of the four bases that form the structures of DNA and RNA. Adenine is also a component of the molecule adenosine triphosphate (ATP), which is a major energy-providing molecule in the workings of every cell in every living thing.

Yet there is one major flaw with the Miller–Urey model and the attempts of later workers, which has recently raised suspicions concerning the relevance of these findings to a description of how life may have originated on Earth. At the time of the experiment, the scientific establishment were in general agreement that the atmosphere of the early Earth possessed a high concentration of hydrogen and a low concentration of oxygen. This, it now appears, was quite wrong.

Far from possessing an atmosphere comprising almost entirely the molecules used in the Miller–Urey experiment (a mix rich in hydrogen), it now appears that the Earth's atmosphere was never so reducing. This does not mean that the mechanisms Miller, Urey and others have described did not take place – it is quite possible that many of the known products they obtained would also have been produced in less reducing conditions – but it does ring a note of caution to the entire theory of amino-acid production by such chemistry on the early Earth.

One striking adjunct to this is that many of the meteorites found to contain organic materials are rich in the very same amino-acids produced by the laboratory experiments, which to many commentators adds weight to the idea that the basic building-blocks of life could have arrived on Earth around 4 billion years ago and set in motion the entire process leading to you reading this book today.

As I said earlier, whichever process is seen to be correct, we are still left with the fundamental problem of how simple amino-acids could have developed into biological material or simple living matter that evolved to produce a self-sustaining ecosystem – the vast range of living organisms that live on Earth. And the particular difficulty lies not so much with the progression from amino-acids through proteins to large molecules such as RNA (although this convoluted series of steps should certainly not be sniffed at); it is the ocean of difference between a 'non-living' or prebiotic molecule, such as RNA, and a system that can partake in the evolutionary process, such as a bacterium. And it is this frontier that is drawing the most avid attention of the biochemists and evolutionary biologists searching for a fuller description of how life evolved on Earth and perhaps elsewhere.

At present there are two theories that attempt to explain how the transition from a prebiotic to a biological system occurred. The first of these is the 'RNA-world hypothesis'. This was first postulated in the late 1960s by biochemists Carl Woese, Francis

Crick and Leslie Orgel. They, like other researchers of the time, realised the fundamental paradox facing anyone attempting to explain how life originated from a soup of biochemicals – that RNA and DNA are needed to make proteins, and proteins are only made if the correct nucleotide sequence is present in the molecules of DNA and RNA. Either we must assume that proteins and DNA or RNA were produced simultaneously and in the same places on the early Earth (which would seem extremely unlikely), or there must be some mechanism to explain how one or the other came first and led to the creation of life.

Their answer to the problem was that early in the process some as-yet unknown mechanism allowed for the formation of a small quantity of a type of RNA. This, they speculated, had the ability to perform a number of roles in addition to the functions it demonstrates today. The RNA that is postulated would have been able to replicate (make copies of itself) without the presence of protein (presumably it would use some of the protein within its own structure) and would also be able to catalyse every step of the protein-production process.

At first this seems more than a little unlikely. We saw earlier that the processes currently occurring in every cell of every living thing are incredibly complex, and require a number of important and diverse organic molecules to play their part. The system employs DNA, RNA and proteins which not only provide the raw materials of the process but also act as catalysts (performed by the proteins known as enzymes). Surely it is asking rather a lot of even a vastly complex molecule such as RNA to perform these other tasks as well as the ones it masters currently.

However, a great deal of work has gone into this area of research since the idea was first proposed some thirty years ago. In 1983, Thomas Cech, working at the University of Colorado, and Sidney Altman at Yale, independently discovered a group of molecules they called 'ribozymes' – RNA catalysts or enzymes made of RNA (hence *ribo-* from ribonucleic acid or RNA and *-zymes* for enzymes). But, although this has given the supporters of the RNA-world hypothesis a boost, there still appears to be no trace of a type of RNA which could have been able to perform the other essential task it no longer performs – self-replication.

While biochemists are hot on the trail of these odd molecules (molecules which would perform the task usually performed by DNA), others are attempting to find ways around the problem. The general idea seems to be that we should not think of the types of RNA, DNA and enzymes present on the early Earth as being quite the same as those present today. In other words, we are encouraged to accept the idea that molecules can undergo some form of evolution.

This might not be as far-fetched as it seems. The underlying principle is that a primitive form of DNA needed a primitive version of RNA, and less elaborate enzymes, in order to catalyse the biochemistry processes of that era. If this was indeed the case, then it is possible that an early form of RNA would have been able to perform several of the jobs that were later taken over by other, better-suited molecules. It has also been speculated that the role of catalyst – a chemical that speeds up a reaction – could have been performed by an inorganic substance within the same environment.*

Because there is still no clear evidence for this mechanism, and no one has yet succeeded in producing self-replicating RNA, the theory remains just that – one of many possible explanations for how life began on Earth.

The other leading contender for a mechanism via which the jump from prebiotic to biological could have occurred is based upon a quite different and highly controversial theory.

Around the time that Crick, Orgel and Woese were formulating their RNA-world concept, another biochemist, A. Graham Cairns-Smith of the University of Glasgow, put forward the rev-

---

* Some readers may be wondering why researchers picked RNA and not DNA as the originator. The reason for this, they claim, is that the ribonucleotides which go to make RNA are easier to synthesise than the deoxyribonucleotides that form DNA. Also, it was easier to visualise how DNA could have evolved from RNA and how it took on its present role as guardian of heredity and the repository of hereditary information.

olutionary notion that the organic agents that led to the formation of living things evolved from inorganic materials.

Initially this might seem rather startling. After all, to most of us, there is a vast difference between organic and inorganic materials. All living things are organic, so of course is all the food we consume. The fields, the trees, the animals and plants that fill the planet are all organic. Inorganic materials include the rocks and stones, the gases that constitute our atmosphere, things generally seen as 'inanimate'.

Although carbon clearly has some exceptional properties, to the chemist it is really just one of the hundred-plus elements in the periodic table, and the distinction between organic (generally associated with 'life') and inorganic (invariably linked with the inanimate) is really a question of perception.

Cairns-Smith points out that, like the biochemical system using DNA and RNA, complex inorganic systems are capable of replicating and passing on information, albeit in a much simpler way. If we hark back to the system that operates in the modern biosphere, DNA carries a code in the form of an almost unimaginably complex set of instructions that is the blueprint for reproduction. RNA and the proteins that have so far featured so heavily in this discussion play their respective roles in bringing this about. What Cairns-Smith proposes is that, around 4 billion years ago, a simpler system operated which initially did not need RNA, DNA or even proteins.

In his system (described beautifully in one of the best popular science books ever written, *Seven Clues to the Origin of Life*), Cairns-Smith describes a process via which two distinct sets of evolutionary path have been followed by life on Earth. The first step was to produce what he calls a 'low-tech' set of machinery using crystal structures present in clays, which would have been abundant at the time (as they are now). These clays, although far less complex than a DNA molecule, can create a self-replication system in which information is passed on from one 'layer' to another, mirroring the way the replication of DNA occurs.

From this low-tech start, Cairns-Smith believes, a gradually more complex system evolved which incorporated organic

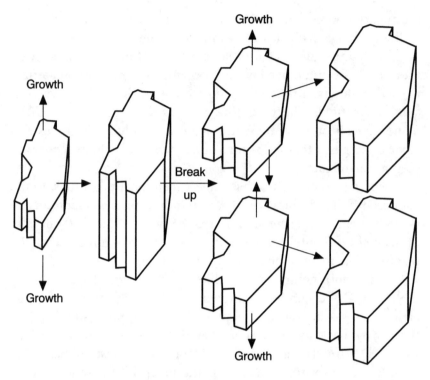

Fig. 6: The way clays can 'replicate'.

molecules. These could have been produced on the early Earth, as the experiments of Miller, Urey and others have implied, or perhaps they arrived here from other worlds upon which life may have already gained a foothold. These unsophisticated systems developed over time into the 'high-tech' machinery we have today, in which DNA, RNA and proteins facilitate genetics, which allows for the evolution of living things via natural selection. He calls the transition point between the low-tech and the high-tech systems the production of gene-1, which then acted as the original template for future development on Earth. This theory goes beyond Darwin, who placed our mutual origins in 'some warm little pool', to a common ancestor that was not even organic.

Cairns-Smith's idea is a bold one and remains on the margins of accepted science, but really it is no more far-fetched than the RNA-world hypothesis, and has a comparable body of evidence to support it. And, linked to both these processes is a final consideration concerning the impetus towards 'life'.

The fossil record shows us that life began on Earth at the very earliest opportunity. In 1980, fossils of creatures called stromatolites or 'living rocks' were found in the Australian desert which for the past sixteen years have been considered as probably the most ancient form of life on Earth, living some 3.5 billion years ago. But, at the end of 1996, scientists working at Scripps Institution of Oceanography in San Diego pushed this date back even further when they isolated a mixture of carbon isotopes* which they claim could have only been produced by living things no less than 3.85 billion years ago. We know the environmental conditions that could have allowed life to form only arose a little over 4 billion years ago, so it would seem that a period of perhaps only a few hundred million years passed before the very simplest life-forms began to appear.

Of course, none of this provides evidence for the evolution of life on planets other than the Earth, but it illustrates the notion that, given the correct conditions, life will appear readily. Carl Sagan once wrote: 'The available evidence strongly suggests that the origin of life should occur given the initial conditions and a billion years of evolutionary time. The origin of life on suitable planets seems built into the chemistry of the Universe.' What he was describing here is the principle of self-organisation.

It has been suggested recently that certain physical and chemical systems can leap spontaneously from relatively simple states to ones of greater complexity or organisation. This organising principle, some argue, is a form of 'anti-entropy' (a reversal of the process which allows order to descend into disorder), and may be in some mysterious way linked to life itself.

Entropy is the 'level of disorder' in a system, and in nature it always increases – fruit left to stand will gradually decompose, its

---

* Almost all elements come in at least two isotopic forms – versions of the atom in which the number of protons (the atomic number) is the same, but the number of neutrons is different, giving the isotope a different atomic weight (the sum of the protons and neutrons in the nucleus).

cells break down and the 'neat', 'organised' form of the fresh fruit will decay into a disorganised mess. The self-organisation principle could, it is believed, help to reverse the natural tendency for entropy to increase in the Universe. Consequently, the chances of life deriving from a collection of complex organic molecules which may have been seeded by simple molecules from space or developed from a 'low-tech' inorganic system is greatly increased.

Even more fundamental arguments are now raging within the physics and biology communities, thanks to the latest suggestions of physicist Professor Lee Smolin of the Pennsylvania State University. Smolin has stirred up controversy by suggesting that universes act like living things by follow Darwinian rules and evolving via natural selection. He calls this principle 'cosmological selection', and according to this theory our Universe is the latest in a line of universes that have been evolving since time began. Furthermore, life in our Universe is merely a by-product of black holes, and it is the creation of black holes that is the motivating force in this Darwinian system. The reason for this, says Smolin, is that the more black holes a Universe possesses, the more stable it is.

But, most remarkable of all, the theory demonstrates that black holes can only form in universes in which carbon is present. According to cosmological selection, the fact that life also depends upon carbon is merely a fortuitous coincidence.[3]

Clearly, we are still a long way from a definitive solution to the origin question, even on Earth. However, if we do soon have the opportunity and the great good fortune to be able to study alien life, this will give us a far better chance of unravelling the complex issues involved in explaining how simple life gained a foothold here and elsewhere. Whichever theory describing the details of how life began wins out, whether it is one of those described in this chapter or some hitherto unrealised hypothesis, evidently a great deal more work is needed and some clear proof gathered to convert what at the moment are only appealing ideas into realistic explanations for what remains one of the greatest mysteries of science.

3

BUG-EYED MONSTERS AND LITTLE GREEN MEN

'It's life, Jim, but not as we know it'

FROM *STAR TREK*

Imagine a Universe teeming with life. Suppose for a moment the sceptics are entirely wrong and that life flourishes throughout our galaxy and others. What then would aliens look like?

There are two distinct schools of thought on this subject. There are those who believe that life on an alien world would have the same level of diversity as it does on Earth, but that the dominant species would look entirely different from us. The other line of reasoning is that most alien worlds would demonstrate a vast array of different life-forms, but that any species advanced enough to create a civilisation would approximate to the humanoid form.

The arguments offered by both camps are complex, and hinge primarily on different interpretations of developmental biology and opposing views concerning the way evolution works on a universal scale.

As we have seen, the theory of evolution via natural selection was made popular by the publication in 1859 of Charles Darwin's *On the Origin of Species*, a work now seen to be as revolutionary in its way as Einstein's theory of relativity, Cubism or Schoenberg's twelve-note scale. And, as happens with all revolutionary developments, the theory of evolution was misunderstood and misinterpreted by people at the time, and continues to be misunderstood today.

During the latter half of the nineteenth century, when Darwin's close friend, Thomas Huxley (nicknamed Darwin's Bulldog) was proselytising the theory of evolution, the popular press and the lay public adopted the completely false belief that 'we are descended from monkeys'. This is a gross misinterpretation of a far more subtle concept. But even today, many people still hold this ridiculous view, and other, slightly more sophisticated misunderstandings have crept in.

In a recent article about life on other planets in a popular science magazine, I was surprised to read that the chances of another planet producing the same genetic sequence as ourselves is placed at 5×10 to the power of 16,557,000 (or 5 with over 16 million zeroes after it). This is of course an unimaginably large number, and suggests that the chances of life similar to that on Earth evolving elsewhere is absurdly improbable. Actually, this statement is entirely misleading, because it fails to take account of evolution.

Indeed, in a universe in which evolution did not occur, the chances of a similar life-form to *Homo sapiens* appearing *is* this unlikely, but genes evolve and mutate according to the environment, and it is likely that the characteristics of alien genes on a variety of life-supporting planets would converge over long periods of time to something not entirely different from those which exist on Earth, assuming the extraterrestrial environment is not totally different.

But Darwinian evolution is even more important than this. It is not simply a mechanism to describe how life-forms develop or how more advanced creatures derive from simpler forms; it is nothing less than the engine of life. Despite the claims of religious bigots, without evolution, there can be no life.

So what is the mechanism of evolution, and how does it influence the argument about alien characteristics? Does evolution give us civilised beings that are bug-eyed monsters (known in the trade as 'BEMs'), does it produce little green men ('LGM'), or does it suggest neither of these?

Evolution is a callous, uncaring process; it is the rock core, the existential nightmare, the dread thing at the bottom of the pit, the fear that keeps you awake at 4 A.M. Evolution via natural selection is the mechanism by which any and all species have their present form, based upon a movement from a less well adapted set of characteristics to another set more suited to its environment. This arc is driven by adversity, by changes in the environment, and by competition (hence another greatly misinterpreted idea – 'the survival of the fittest'). The greater the degree of adversity an organism experiences within its environment, the faster the rate of evolution.

The reason I say that Darwinian evolution – or evolution via natural selection – is cold and uncaring is because it is purely mechanistic: the changes wrought by evolution are not guided or governed, there is no God to pull the strings and mastermind the process.

Darwinian evolution does not need God; it is a self-perpetuating process and works at a genetic level. Those members of a species who are best adapted to their environment will have a greater chance of survival and therefore a greater chance of breeding successfully. Through sexual reproduction they pass on their genetic material, which crucially is also scrambled by the reproductive process and subject to chance mutation. These mutated characteristics (along with others that move through the reproductive process unchanged) will be found in the subject's offspring. If they have received genetic material that gives them an advantage within their environment, they too will have a greater chance of survival and will be more likely to breed. So, over many generations, characteristics will be inherited which make members of the species better adapted to their environment. Chance mutations, which are purely a matter of serendipity, will also mould the species, and those specimens that draw the evolutionary short straw will pass unloved into history.

This mechanism is almost certainly a universal truth. Again, I run the risk of being accused of chauvinism, but as with my argument in the last chapter concerning the uniqueness of carbon, the mechanism of evolution via natural selection is a fundamental law and there is no reason why it should not operate throughout the Universe. As I said earlier, without evolution there can be no life as we would recognise it.*

So how does this very simple mechanism determine the way we look or the way a whale differs from a canary?

The discipline of developmental biology is really a blend of several aligned fields, including evolutionary biology, palaeontology and genetics, which looks at the way in which creatures on Earth have evolved from simple forms dating back billions of years to the fauna and flora we witness today, and of which we are a part.

Developmental biologists start from two basic foundations. One path, called evolutionary biology, is the study of the range of modern animal structures, which have been called the 'body plans' of animals – the fundamental groupings of different animal types we see all around us. From these it is possible to work backwards using computer models to determine from where these basic plans derived.

The other approach is 'genetic retracing', which involves analysing how genetic material has changed over long periods. Genetic characteristics are of course one of the key factors in determining the nature and diversity of life, and it is possible to trace back the way genetic material has altered both within species and across different species. In this way, biologists can draw conclusions about the nature of a common ancestor of two or more modern species, a process which can then be pushed further and further back to generate a 'family tree' of species.

Neither of these techniques is straightforward. The retracing of physiological forms requires powerful computer models and relies on a vast array of parameters, information acquired from palaeontological and archaeological finds, how function ties in with design,

---

* Having said that, even proposed life-forms as strange as intelligent gas clouds had to evolve to develop intelligence, and may have developed from a different gas cloud that was less well adapted to its environment.

and improvements in our understanding of how the physiology of animals relates to environment. Genetic retracing is problematic because different genes, like different species, evolve at different rates, and the evolutionary lines can branch out in a vastly complex network leading researchers along blind alleys and up logistical culs-de-sac. But gradually the developmental biologists are constructing a picture of how life evolved on Earth which might lead to conclusions about how life could have also developed on other worlds.

As we saw in the last chapter, life appears to have gained a foothold on Earth around 3.85 billion years ago. Quite how the transition was made from prebiotic material to what we would define as a 'living' organism is still a matter of great debate, but clearly, after that transition point, life flourished and evolved. But it would be a mistake to think that this was a simple linear progression.

Until as recently as about 530 million years ago, the most complex form of life on Earth was an organism no more advanced than simple algae. Even then algae were a relatively recent arrival. For the first 2.85 billion years in which life could be found on Earth, it consisted of single-celled organisms such as bacteria. Then, about 1 billion years ago, the first very simple multi-cellular creatures, the algae, appeared for the first time. Later, around 550 million years ago, during a period geologists call the Neoproterozoic era, an array of more advanced (but still relatively primitive) organisms made an appearance. These simple creatures probably resembled modern-day sea pens, jellyfish, primitive worms and slug-like animals all of which have left faint fossil remains and markings.

But then, some 20–25 million years later, there was a complete transformation in the evolutionary development of life on Earth. Throughout the Neoproterozoic period, the Earth had been populated by animals which displayed a relatively small collection of different body plans, but suddenly everything changed. If we compare the fossil record before 550 million years ago to that around 520 million years ago, there is a dramatic difference in the number and types of organism living on planet Earth.

This sudden change is called the Cambrian explosion, a burst of activity in the life of the fauna of this planet (an era that bridges the Precambrian era, extending from about 4 billion years ago, and the Cambrian period, which lasted until around 500 million years ago). If an alien visitor had arrived on Earth before this upheaval they would have found scant evidence of life other than bacteria (which would, incidentally, have been as prolific as they are now). But if they had arrived during the Cambrian period, immediately after the Cambrian explosion, they would have observed an array of different organisms. And it is these primitive beings, including almost every known type of shelled invertebrates (clams, snails and arthropods), appearing at this time that gave rise to the modern vertebrates – including, of course, *Homo sapiens.*

Furthermore, by the close of the Cambrian explosion (a period which probably lasted only a few million years), all the basic architectures, the body plans of all animals on Earth, had been established. From that point on, all evolutionary steps (including what we consider one of the most dramatic – the point at which some animals left the sea to live on land) merely required subtle refinements of the basic animal types established during the Cambrian explosion.

From these foundations, animal life on Earth has arrived at just thirty-seven distinct body plans which encompass the entire hierarchy of animals upon which the taxonomic classification is based.* Utilising just these thirty-seven body plans, nature has offered up a vast diversity of creatures ranging from a centipede to Albert Einstein.

Although on Earth one animal looks very different from another, on a genetic level there is a surprising degree of compatibility across species. Naturally, all life on Earth is carbon-based and all life here relies upon DNA to carry the genetic code, but

---

* The system used by biologists for filing the animal and plant world into more and more inclusive categories. Closely related species are grouped together into a 'genus'. Genera with similar characteristics and origins are grouped into 'families', families into 'orders', orders into 'classes', and classes into 'phyla' (in the case of animals) and 'divisions' (in the case of plants). Finally, related phyla or divisions are placed together into 'kingdoms'.

genetic consistency goes beyond these fundamentals. Geneticists have discovered that, at their core, almost all living things share a collection of genes called 'regulatory genes'. These genes determine the essential body plan of a creature and, because many groups of animals share a common ancestor, these genes are very similar in creatures we think of as being very different, such as, say, a squid, a zebra and a fly.

Most animals start from a single cell – the fertilised egg or 'zygote'. This cell then divides and multiplies, proteins are produced which have specific roles – forming organs, glands, the skin, muscle and bones of the organism. But at the core of each organism is a set of common genes which control the process of protein formation. These proteins then interact with other genes which produce more proteins, and so it continues in what is called a 'gene cascade'. As this process continues, the genes become more and more specialised and differentiated from the central core of common genes at the heart of almost all animals on the planet.

The simplest instruction for an embryo during development is that which determines its own body axis – which end becomes the head end and which the tail; which develops into the back and which becomes the front. This instruction comes from a collection of genes at the heart of the cascade, and is an example of a shared characteristic between almost all species. Further down the cascade are genes which determine whether a head is developed

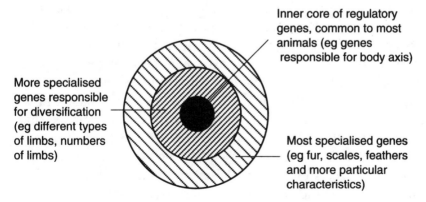

Fig. 7: A schematic representation of the regulatory genes in animals.

outside the trunk of the body, and genes which control the development and growth of limbs. In species as diverse as sting-rays and horses, for example, these will be different, but they will be very similar for horses and sheep, or even horses and flies. It is only further down the cascade that there is a marked differentiation between very different creatures. Even birds' wings and the forelimbs of a mammal can be thought of as being controlled by a similar set of regulatory genes.

The origin of this process lies with a sequence of DNA which is known to be truly ancient. Called the 'homobox', it is thought to predate the origin of animals on Earth. In other words, this DNA sequence – a vast and highly complex collection of base pairs – existed during the Precambrian period. A great deal of research has been done in this area of genetics during the past few years, and of particular interest has been a set of genes containing the homobox DNA sequence called 'Hox genes'. These are normally found clustered together in animal chromosomes, from which they derive the name 'Hox clusters'.

Recently, scientists studying these Hox clusters have come up with some truly amazing results, which show that these genes are almost like templates for the animal of which they are a part – the genes are arranged in the cluster in the same way the animal part they control is positioned in the growing embryo.

Geneticists and evolutionary biologists use Hox gene clusters to help explain how different species may have shared a common ancestor before the Cambrian explosion, and how the evolutionary paths which have led to modern species were arrived at. For example, they can show that the mouse and the fly had a common ancestor before the Cambrian explosion, because the regulatory gene which initiates the growth of the eyes in the mouse is so similar to that in the fly that the genes can be interchanged in the respective embryos without any adverse effects on their development. Naturally, the more specialised genes which produce quite different *types* of eye in the fly and the mouse could not be interchanged successfully, but these are genes which evolved differently after the common ancestor brought forth two distinct species after the Cambrian explosion.

The mouse embryo

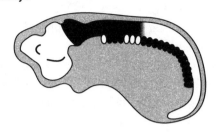

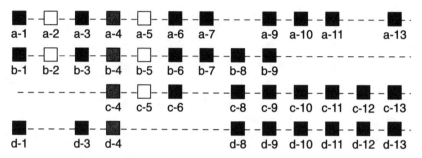

The Hox gene cluster of a mouse showing how the genes of a mouse are
arranged in the same way as the functions and anatomical characteristics for
which they are responsible

Fig. 8: The arrangement of the genes in a Hox gene cluster compared
to the anatomy of the embryo in which they are found.

Obviously, the closer two species are, the greater the genetic
comparisons and the more regulatory or core genes they share.
This is particularly striking when we consider the genetic make-
up of the primates, which reveals that there is only a 1 per cent
difference between the genetic structure of a human and a chim-
panzee.

So how does all this relate to the possible nature of life-forms
on other planets – the science of exobiology?

The answer lies with the fundamental fact we have already
considered: that all life as we would recognise it is based upon
carbon. Carbon and the compounds it creates lie at the core of
the 'molecule of life', DNA, and DNA is the key to genetics,

which has provided the diversity of creatures on Earth. It could be argued that there are alternatives to DNA; that there is perhaps a molecule that does an even better job. This is just about conceivable, and if it has happened elsewhere then an entirely different set of genetic material might have been created which could have produced an altogether alien set of creatures on other worlds. These would still be carbon-based life-forms but, because their genetic structure would be based upon a different collection of molecules, they may end up looking very different from anything we have ever seen. But is this scenario actually possible?

One interesting concept that has been taken seriously only during recent years is the idea of 'convergence'. This notion takes on many different forms in fields as disparate as biology and engineering. Put simply, the idea can be expressed as: 'Many very different starting points produce a limited number of solutions to a problem.'

An everyday example is the car. The 'problem' in this example is to produce an efficient vehicle for a reasonable price which will move a small group of human beings from A to B at a reasonable speed and in relative comfort. Now, if one had never seen a modern car, you might be forgiven for thinking that there would be a vast range of ways in which this could be done, but really there is not. Certainly, we have a collection of basic types of vehicle (which we may think of as being analogous to body plans in the animal kingdom) which, on the surface, look quite different from one another, such as coaches, cars, buses and lorries. But the differences between them are all relatively superficial and come down to cosmetic design, size, and fairly subtle refinements in their shape. At their root, they are all metal structures with doors, wheels, roofs, cabins, steering wheels and petrol tanks; they all use fossil fuels; they have a front and a back and seats on which humans sit and they all travel along roads.

If we were to build a factory which manufactured all the types of vehicle on the roads of the world it would have to be a vast and elaborate plant (analogous to the ecosystem of the Earth in which all creatures are produced), but it would only need to manufacture a relatively small number of basic body shapes, and the manufacturing process would have many shared basic components (analogous to the growth of body parts controlled by the regulatory genes). But, most importantly, we humans have arrived at a solution to this problem of travelling by creating vehicles that look very similar.

In the same way, nature deals with the problems it faces in a limited number of different ways, and it always does this in the most efficient way possible – which leads on to the idea that life may have developed on other worlds by using a different complex molecule to DNA. However, it might be argued that DNA is the best option. Using DNA, nature can solve the problem of reproduction and at the same time allow for evolution via natural selection which relies on the ability of the genetic material to mutate from one generation to the next. There might only be one system in the Universe for doing this, and nature operating on Earth has achieved this 'perfectibility' over billions of years of experimentation.

So, if we assume that DNA is the best molecule to use for passing on genetic information, we are led to the conclusion that a small group of common ancestors would have evolved on any planet where there is life, and that these common ancestors would have developed into a range of creatures that adapted to their environment. But what would that environment be, and how would it influence the nature of alien species?

I have mentioned elsewhere that bacteria have been found in some extreme environments on Earth – in radioactive waste, thousands of feet beneath the sea-bed, in hot springs and in the frozen wastes of Antarctica – but bacteria are very hardy; which, incidentally, is why they are, in purely biochemical terms, the most successful living organism we know of. More sophisticated animals could not have evolved from single-celled organisms in environments as harsh as those in which bacteria appear to thrive.

This places limits upon possible environments on alien worlds harbouring any kind of highly developed life-form. First, the temperature of the environment must lie between 0°C and around 40°C (much hotter than this, and enzymes cannot operate and denaturing occurs). The environment must not be flooded with intense radiation, since this damages biochemicals and inhibits many of the chemical reactions required for mechanisms which control the functioning and continued growth of cells. A range of atmospheric pressure presents less of a problem, but will, as we shall see, have a great influence upon the physical characteristics of successful life-forms. And finally, of course, the atmosphere must have the correct balance of oxygen, nitrogen, water vapour and carbon dioxide to allow for any ecosystem consisting of flora and fauna based upon DNA.

Beyond these fundamental considerations are more subtle requirements which could hold the key to whether advanced life-forms evolve commonly on other worlds. Previously I described how there was an explosion of life on Earth around 530 million years ago – the Cambrian explosion – and that until this point in the history of the Earth the most complex organisms on the planet were a few varieties of simple multi-celled organisms such as primitive algae. This relatively barren period preceding the Cambrian explosion lasted around 3.3 billion years, some 90 per cent of the time in which life has existed on Earth. So what was it that precipitated the Cambrian explosion? What pushed life on Earth on to a new level, initiating the process that brought forth the range of organisms alive today?

Nobody really knows the answer to this question, but it is no exaggeration to say that its solution offers the single most impor-tant factor in establishing the probability of life on other worlds and what that life would be like.

There are two competing theories used to help explain what hap-pened on Earth 530 million years ago. The first is simply that the time was propitious, that nature had experimented with various evolutionary mechanisms and eventually hit on the right way for-ward. This is not to suggest that nature could 'see into the future' or was 'guided' by some plan that would end in the production of a dominant species like *Homo sapiens* – natural selection does not operate by any sort of plan, it is driven by success and buffeted by random events. Rather, in one sense, this explanation of the

Cambrian explosion is an adjunct to the idea of convergence – nature had found a path through the maze. It had taken over 3 billion years to reach this point, but it worked eventually.

This explanation is a comforting one for those who believe that life flourishes beyond our little world. If such a process could have taken place here by its own volition, then it is perhaps an example of a fundamental process – a consequence of life reaching a certain level of complexity, a point from where it is spurred on to the next stage.

The rival to this view is altogether more pessimistic, and suggests that the Cambrian explosion was precipitated by some unknown freak ecological event. Contenders for this range from comets or huge asteroids colliding with the Earth to sudden dramatic changes in the level of oxygen in the atmosphere.

It is known that, like all the planets in the solar system, the Earth has been bombarded by meteorites, comets and asteroids throughout its history, so it is quite possible that one such incident occurred at just the right time to precipitate the Cambrian explosion. Such an event could have so disturbed the ecology of the planet that evolution was given a fresh impetus.

By the same token, changes in atmospheric conditions cannot be ruled out. An increase in the percentage of oxygen in the atmosphere would almost certainly have triggered a dramatic surge in biological activity on the surface, and could explain why a vast array of fresh life-forms appeared during such a brief period.

If any of these explanations are true, it means one of two things. Either these events did in some way encourage a blossoming of life, but that it would have happened anyway; or that one of these global changes did occur and is solely responsible for sparking the Cambrian explosion. If true, this latter possibility would mean that advanced life-forms, ones that would have a chance of evolving into creatures that could create a civilisation and technology, are far less common in the Universe than many would like to hope.

Having considered the pessimistic option, let us return again to an optimistic line and assume for our purposes that the Cambrian explosion on Earth was due to a natural mechanism, a system that is in some way linked to the development of life once it gains a foothold. What then can be said about the possible ways in which advanced life-forms developed? Would evolutionary routes on other planets lead to creatures similar to those seen on Earth, or would they be utterly different?

Once again, opinion on this subject is split between those who suggest that an anthropomorphic line might have been followed by extraterrestrial organisms and those who claim that it is ridiculous to believe that creatures similar to mammals could have evolved on distant planets, perhaps thousands of light-years away. Indeed, one of the leading authorities on the subject of exobiology, biologist Jack Cohen, has said: 'It is easy to be drawn along a line of reasoning that proposes that alien organisms developed in similar ways to creatures on Earth, but if we are not careful it leads us to suggest that aliens also drive Ford Capris.'

So what conclusions can we draw? Which arguments can we be confident about, and what remains utterly unpredictable?

First, we can be safe in assuming that any alien life-form will be based on carbon, and we can be fairly sure that it will operate using biochemical mechanisms built around DNA. We can then take one of two views concerning the way it might evolve. We can assume that there are many different solutions to the problems living creatures face in the battle to survive, and that this will bring forth an almost infinite range of animal types, including creatures that do not conform to our definition of life as we know it. The alternative is to adopt the concept of convergence and assume that this is also a universal mechanism and that there are only a limited number of solutions to the demands placed upon living things.*

If we assume convergence does operate universally, we then have to consider the possible range of environments within which life would have a chance of developing into advanced life-forms,

---

* Support for this idea comes from a fact mentioned earlier – that on Earth, there are only thirty-seven different body types, and the millions of species we know of all fit into those categories. There is no reason why this should not also happen on other worlds.

and how they could have evolved. The environment on a particular planet cannot be too harsh or else any form of complex multi-cellular beings could not have evolved. On the other hand, the environment has to offer a challenge to living things so that natural selection can operate and evolution can occur. Throughout this book I should emphasise that I am only interested in considering creatures with which we would have a chance of communicating. Although this does not restrict us to English-speaking humanoids in pin-striped suits, it does, to some extent, limit the environmental conditions that could support any kind of advanced life-form. We must assume then that any planet which brings forth a civilisation must have an ambient temperature comparable to that on Earth, and that the intensity of harmful radiation is manageable.

But what about the atmosphere? It is perhaps too easy to assume that a planet that could be home to an advanced life-form must be rich in oxygen and possess the correct balance of oxygen and carbon dioxide. This is because on Earth we live in a balanced ecosystem, in which plants need carbon dioxide to facilitate photosynthesis which in turn produces oxygen. All animals use oxygen, which is transported by the blood and carried to the cells of the body where it is involved in almost all the biochemical mechanisms that maintain us. But would it not be possible to have a successful ecosystem using other gases?

If we are to consider alien life-forms that have DNA at the core of their biochemistry (and it is difficult to imagine alternatives), then a respiratory system that utilises gases other than oxygen presents problems.

The first of these is that the planet would have to possess a completely alien but fully-integrated ecosystem using gases other than oxygen and carbon dioxide. This is because no alien creature could evolve or remain alive on a planet without interacting with other organisms. In other words, they themselves must be part of an ecosystem, and any ecosystem must include gaseous cycles similar to the oxygen–carbon dioxide cycle on Earth. Such alien systems would have to integrate organisms similar to our plants and animals. An alternative might be a system which operates by a gaseous exchange between animals and some form of rock that processes gases in a similar way to plants, taking up one gas and producing another in a process similar to photosynthesis.

The second problem is this: it may be possible for animals on an alien world to breathe a different *mixture* of the same gases as we do, but is it conceivable that animals based on DNA could use any gas other than oxygen? The only experience we have of such an event is when humans (or any animal) breathe in a toxic gas such as carbon monoxide. Our systems can take only very low doses of foreign gases before the systems in the cells of our bodies are corrupted, causing great physical distress followed rapidly by death via asphyxiation.

Yet this is perhaps a rather unfair comparison, because any alien life-form that used different gases in its biochemical processes would have evolved within such an environment, and so would find it quite natural to breathe, say, carbon dioxide or ammonia. Even so, the original template that used DNA to allow evolution to occur must have operated with very different biochemical processes to allow such a thing, and it is highly speculative that such different mechanisms are possible. Most biochemists would agree that cells based on similar chemistry to ours could not process such gases as nitrogen, hydrogen or hydrogen cyanide – or indeed any gas other than oxygen.

So, what of other environmental considerations? What of atmospheric pressure and gravitational fields?

On a planet where the atmospheric pressure is higher, it is possible that intelligent creatures could have evolved which look very different from humans. The lay-out of the respiratory systems of such creatures would probably be very different because the pressure of the gases they breathe would not be the same as it is here, which means that the processes allowing gases to diffuse into their version of a circulatory system (if that is what they possess) would operate at different potentials. But, once again, nature would have found solutions to these differences and compensated for them in an appropriate way.

More problematic could be the factor of gravity. The strength of the gravitational field on an alien world in part determines the relationship between the body structures and sizes of creatures living on that world. Within relatively narrow parameters, all land mammals on Earth are of a similar size – there are no 200-feet-long quadrupeds, nor are there insect-sized mammals. But what would be the situation on a planet very much larger than Earth, where the force of gravity might be twenty times or a hundred times greater than it is here?* Alternatively, how would life have evolved on a world where the gravitational pull was considerably lower?

Such worlds would certainly produce creatures that were generally larger or smaller than most mammals on Earth, but there are problems with this. Very large creatures need very large hearts to supply the volume of blood needed to keep their bodies functioning, and big hearts need big lungs. As we know from the design of animals on Earth, this is not a strict limitation, but for an animal to evolve into the dominant species and to establish a civilisation, they also need a large brain that is not simply devoted to running a massive body. Such a brain would need a large head and still more blood to supply the cells with oxygen, which needs a huge heart and massive lungs, and so it goes on. This also presents difficulties with the ratio of body parts, mobility and efficiency, which could restrict the evolutionary success of such a species developing within a highly competitive environment.

As a counter to this, it has been said that the dinosaurs were far more successful than humans because they lived on the planet for some 100 million years in one form or another, whereas we have only been 'human' for little more than 1 million years. This statement is, of course, nonsense. Success, even in purely biological terms, is not based merely on longevity, but must take into account the role an animal plays in the ecosystem. Humans are the only animals to have created a civilisation to control their environment in any large-scale way, and the time is fast approaching when our technology will enable us frail animals to control the environment far more effectively. We are also the only animals on

---

* This is in itself a complex matter because, for a number of different reasons, it is likely that solid planets many times larger than Earth could not exist.

Earth with the capacity to leave our world if we wanted to, albeit a limited option currently.

So, within quite strict limits, height and bulk are factors that do not prohibit an animal developing a civilisation. Such differences between 'us' and 'them' are little more than cosmetic, or what Jack Cohen calls 'parochial characteristics'. Whether or not an extra pair of limbs or a third eye would be favoured by an ecosystem on another world is open to debate. There may be advantages in having these things, but in any environment, nature will always go for the most efficient option. If the advantages of a third eye or an extra pair of ears outweighs the demands produced by the extra weight, the extra blood requirements, the development time (both in terms of evolution and within the womb), then it could happen within a different environment to ours. If not, it is unlikely nature will follow that course.

But how far can we go with this argument? If we argue that the environment within which a civilisation can develop cannot be that different from the one we experience on Earth, do we indeed end up suggesting that aliens drive Ford Capris?

There are those who do take the anthropomorphic argument to its limit, and suggest that the most likely design for a successful life-form that has evolved to the point of developing a civilisation will be similar to ours – that they will look like us (or us like them). But how valid is this argument? For example – do we need more than two legs? Those who take the anthropomorphic stance say no. But what if the alien world is always swept by strong winds? The counter-argument could be that such an ecosystem could not exist because it would have been very difficult for plants to have evolved. Do we need two heads? We have two of most things, but as with the argument against bigger heads, two brains would require too much blood for the same-sized heart, and we are back in the same cul-de-sac. But, according to the counter-argument, convergence would not allow for such a consequence anyway, it would enforce the most efficient solution – a biped with one head is better than a large ungainly biped with two. We only have to look at our own planet to see that this final argument is probably right – after all, how many two-headed creatures have you seen when sober?

So far, I have concentrated on the physical characteristics of possible alien beings – the nature of their biochemical systems and how these could affect their appearance. But if we do make the assumption that there are other civilised beings living on distant worlds, what can we predict about their brains and minds?

When dealing with this aspect of exobiology we have even less on which to build a framework of ideas than with the physiological arguments, because, with perhaps the exception of a few other special-case species on Earth, the only animal with any form of conscious social interaction or intellect (as distinct from pure intelligence) is *Homo sapiens*. We are the only animals on the planet to keep records, to have developed a recordable language – a form of writing whether it be Sanskrit or modern English – to have built a civilisation based upon trade, and, crucially, to plan, to have a concept of our place in the world and flow of generations of our species.

So perhaps the first question to ask is: what is it about us that makes us different from the other species on the planet? If we can arrive at an answer to that, we might be able to extend the principle to extraterrestrials.

The difference between us and other species seems to come down to a matter of what is loosely termed 'intelligence', but even this is a little vague. What is intelligence? And are we actually more intelligent than some other animals, such as dolphins?

One of the most important factors is brain size. We have very large brains for our bodies. If the human brain had a covering layer which were to be unfolded and spread out, it would cover four sheets of A4 paper. By comparison, a chimp's brain would cover only a single sheet; the brain of a monkey would just about meet the dimensions of a postcard; and a rat's brain would barely cover a postage stamp. However, it is not just a matter of size. Dolphins have very large brains, yet they have not developed a civilisation, and it is believed that almost all their brain capacity is involved with managing their complex sonar system.

It is undoubtedly true that a relatively large brain certainly seems to be a prerequisite for any species that hopes to dominate their ecosystem and to become the 'rulers' of their world. The reason for this is that a large brain is essential for developing a skill as incredibly complex as language, and language is a basic (but not the only) requirement for civilisation and social development.*

In the case of human history, there was a 'sudden' four-fold increase in brain size between 1.5 and 2.5 million years ago. It is thought that before this point, early human ancestors had a brain capacity comparable to that of a chimpanzee. What caused this rapid development remains a mystery, but it marks another key turning-point in the development of the human race. The most likely explanation is that our ancestors were faced with a severe 'challenge' to their continued existence. The best candidate for this is the advent of the most recent Ice Age, the Quaternary Ice Age, which is believed to have acted as a 'filter' for many species, including *Homo sapiens*.

In an example of nature applying the maxim 'What doesn't kill you makes you stronger', it would seem that early *Homo erectus* (the immediate progenitors of the first *Homo sapiens*) learned a great deal from their experiences in colder climes. Biologists have reached the conclusion that the more intelligent land-based animals are omnivorous. This they believe is because omnivores can adapt their tastes to find diverse sources of food, and the effort to search out new resources is also a learning process which helps the animal develop skills not common in purely carnivorous or vegetarian animals. Similarly, the increase in brain capacity precipitated by the Ice Age came about because of the demands this placed on early humans. Those who survived this change in the environment did so by learning to find and use new resources, which led them to gradually develop social skills, to create communities, to develop language and to eventually take the first steps towards civilisation.

And with language comes what we call 'intelligence'. If we define intelligence as the ability to communicate and process

---

* It has been suggested that an alien civilisation may not use language but a form of telepathic communication, but the development of this too would require a large brain.

ideas, then a huge leap in human evolution came about with the development of syntax, and from that the ability to string together meaningless sounds (phonemes) to make 'meaningful' words. This ability enables us to create sentences and to communicate abstract ideas, to plan, to create social rules, taboos and hierarchies. Language really is the cornerstone of civilisation.

But would there necessarily have been events comparable to the Ice Age on other worlds? It would seem very likely. Although there are a number of plausible theories to choose from, nobody knows for sure why the series of Ice Ages occurred on Earth and whether these are linked to very common natural processes in the life of a planet. But it would seem reasonable to suppose they are common to a good percentage of worlds. And, of course, Ice Ages may not be the only form of challenge an embryonic dominate species might be offered – other worlds may suffer environmental changes precipitated by comet or asteroid collisions, volcanic activity or short-lived irregularities in the behaviour of the planet's sun (although, naturally, none of these could be so severe as to destroy the embryonic 'master race').

So, the chances are that if a planet were able to sustain life, then that life would evolve to the point where a dominant species emerged which would then be faced by a challenging situation (or perhaps several such situations) which would nudge that species along an evolutionary path that would eventually lead to the development of a civilisation. But what form could alien civilisations take?

With this, we really are entering the realms of speculation, but we can at least lay down a few guidelines. Earlier, I said that with the exception of a few special-case species, *Homo sapiens* was the only creature on Earth with any form of intellectual or social interaction. To complete our discussion of how 'biochemical' alien minds might work and how extraterrestrial societies could operate, let us now look at those exceptions.

Many people tend to think of dolphins as highly intelligent creatures, and they almost certainly are (by some definitions), but they exhibit a form of intelligence that appears to be very different from ours, one which has not led them to the creation of what we understand by a 'civilisation' or a 'society'. Why is this?

The simple truth seems to be that dolphins did not have a chance of competing with humans because they live in an environment that makes it very difficult for an intelligent animal to create any form of infrastructure. And this is due to several complex factors. First, the other part of the developmental formula to complement brain capacity is the need for physical versatility. Dolphins do not have digits with which they can manipulate materials, and they certainly have not developed opposable thumbs, which have been one of the most important distinctions between human and non-human primate development on Earth. Dolphins are a very successful species – their physiology has evolved in a way that allows the animal to be perfectly adapted to its environment, but they were unable to even start on the road to civilisation.

Dolphin 'language', although sophisticated compared to almost all other animals on the planet, has not developed in a way that can lead to social development beyond a rudimentary level. With the bodies they possess, they could not have constructed the aquatic equivalent to buildings; they could not easily record any knowledge they acquire in the way humans have via the invention of writing; they could not easily cultivate their territory or manage other animals, which means they are constantly at the whim of fluctuations in food supply. Finally, they could not have developed weapons, so they could not have engaged in one of the most important 'civilisation-building' activities – the waging of wars. In short, it seems that the odds are stacked against any intelligent aquatic animal developing a civilisation as we understand the term.

Any planet that brings forth life, therefore, must have enough land to allow animals to develop, and for the correct ratio of plants and animals to arise to create a balanced ecosystem.* Furthermore, any world which is entirely covered in water will almost certainly

---

* Although, in theory, it might be possible for a special system to emerge and flourish on a world almost entirely covered in water which contains animals and water-based plants rather than land-based plants.

not harbour life any more advanced than the equivalent to Earth-style fish and relatively simple aquatic animals and plants.

A final alternative form of biochemical life that could be viewed as 'intelligent' is the 'collective organism'. On Earth, there are only a few examples of what could be loosely put into this category; the most obvious examples are bees and ants. These creatures are individuals, but each bee or ant merges, at least in part, with the hive or the colony. In the case of ants, such colonies can number up to 20 million individuals.

Once again, bees and ants are extremely successful species, but despite the best attempts of 1970s Hollywood film directors, we do not see them as being the dominant species on this planet. This might be simple vanity on our part, or it could be justified (only time will tell); but, if we encountered an alien life-form on another world which acted as a single organism by the merging of millions of individuals, would we recognise it as an extraterrestrial civilisation?

Perhaps a more pertinent question is: could such a civilisation evolve anyway? Because we know so little about the behaviour of ants and bees on Earth, it is almost impossible to answer this conclusively, but we might assume with some justification that one of the reasons ants or bees have not formed a super-organism with the characteristics of a civilisation is that communication between individuals is not sophisticated enough. On our particular planet, evolutionary factors have not favoured the emergence of such a dominant civilised species. However, that is not to say it could not occur if the conditions were appropriate elsewhere.

Finally, before we leave the subject of biochemical life-forms altogether, there is one further concept to consider. Could it be that we humans are ourselves following a road to the formation of a super-organism? I began Chapter 2 with the question 'What is life?', and posited the idea that the Internet might be considered to be 'alive'. Is it not possible that, as we develop an increasingly elaborate 'technological web' (a conglomeration of infrastructures centred primarily around communications and the sharing of information), we are turning into a super-organism? Once again, this is another matter that will only become clear with time, but to conclude this round-up of possible life-forms on other worlds we should explore the possibility that artificial intelligence has been developed on other worlds.

Artificial intelligence is a hot subject at present, and teams of computer engineers, programmers and pure mathematicians around the world are trying to develop a system that thinks for itself. John Taylor, Professor of Mathematics at King's College, London, once famous for his investigations of spoon-bending, is now more excited by computer simulations of the brain which he uses to explore the meaning of consciousness – the physics of how we think. Meanwhile, Francis Crick, co-discoverer of the structure of DNA, is independently using complex mathematical modelling to help unravel the foundations of human self-awareness and the mind.

These simulations are called 'neural networks' and are really very simple models describing how parts of the cerebral cortex (the outer, folded part of the brain) operate – an example of silicon hardware made to duplicate living 'wetware'. Taylor, Crick and others believe this line of research could one day lead to the manufacture of intelligent machines, ones that learn, as we do.

Neurophysiologists have identified a part of the cerebral cortex called the *nucleus reticularis thalami*, which acts as a 'gate-keeper' for signals in and out of the region where higher thought processes occur. Very simple versions of this have been produced using sophisticated computer programmes, and it is now becoming clear that 'thought' is a result of a tussle for supremacy conducted by many competing impulses, drives and memories. Whichever neural activity wins the battle, it then broadcasts itself across the whole cerebral cortex as a 'thought'. By using neural networks, it might be possible to piece together the convoluted steps from impulse to 'conclusion', stimulation to 'understanding'.

In the cerebral cortex of the human brain, thinking is facilitated by about 6 billion cells called neurones, which produce a neural network containing countless trillions of connections. These neurones are grouped together in six layers. In each layer, around a thousand cells form a column and there are over a million columns. In current simulations, the computer deals with many magnitudes fewer units, but even with these, they can be taught to respond intelligently to very simple stimuli.

For example, a three-layer system can learn to recognise a figure, for example a '£' sign. Using a device that acts like a simple retina, the image is made to stimulate the first layer of units. These impulses are then processed by the next layer, and if they recognise the binary code representing the figure, they pass a signal on to the output layer. Success at recognition means the connections between the layers of units is strengthened electronically, so it responds faster next time it receives the same signals. In itself, this is of limited use, but when other networks working in parallel have connections strengthened, the whole system can be said to gradually 'learn' something.

And this is exactly how we learn – through a step-wise process but linked in parallel. This mechanism accounts for the huge difference between the way we think and the way even the most sophisticated conventional computer operates. In human brains, each sequence is conducted in parallel with millions of others simultaneously, but almost all computers outside of research labs operate in a completely linear fashion, solving one simple problem (albeit very, very quickly) before moving on to the next. Because of this, the *intelligence* of living things and the *cleverness* of computers are very different, and it means that a crayfish is far more intelligent than a Cray supercomputer simply because it can learn. A machine that operates by linear processes could never develop any form of consciousness, but if neural networks could be made complex enough they might eventually develop into artificial brains.

So, how likely is it that we will one day develop genuine artificial intelligence? And is it an inevitable step in the evolution of technological societies that may exist throughout the Universe?

René Descartes famously wrote 'I think, therefore I am', but there are some who suggest that he had it wrong, and that his famous pronouncement should have been 'I am, therefore I think'. The reason is that learning is not all there is to consciousness. Humans – and perhaps some other animals on this planet – appear to go much further. Our brains have become so sophisticated we have achieved a form of *Gestalt*, a level of complexity that produces something greater than the sum of its parts – something we label 'self-awareness'.

Could it be that a sufficiently complex neural network might be made to acquire this same semi-mystical quality? And could such a sophisticated 'life-form' become self-sustaining and perhaps even dominate the biochemical species that had created them in the first place? This scenario is reminiscent of a thousand and one science-fiction films, and is rooted perhaps in the primal fears stirred up by such tales as Mary Shelley's *Frankenstein*, but there is no reason to doubt that a sufficiently old and technologically advanced civilisation could not create intelligent, independent sentient beings, which may then supersede them.

Another angle on the same argument is the incredible advances we have seen in cybernetics during the past ten years. Already, cyberneticists and computer engineers are talking seriously about the possibility of producing computerised replacement body parts (for both external and internal use) and the ability to transfer human memories and stored information on to microchips. Those working at the cutting edge of this technology visualise a future where, in theory, we need never die but instead replace the old bits of ourselves with new parts and even transfer our minds to new, artificial bodies. Ignoring any ethical hurdles this technology erects, the possibilities are endless, and could have been fully exploited by any civilisation a few thousand years more advanced than us.

All of this leads to what commentators of future technologies describe as 'unnatural selection' – the situation where a species advances to the point where it can manipulate its own biological evolution and even supersede it. A few thousand years is as nothing when considering the time required to evolve species and develop rudimentary civilisations. If we can now visualise doing these things, then any visitors we do encounter will probably have already done them.

To conclude, whatever our personal views may be about the nature of alien life, the truth will only be known when we one day make contact with alien intelligence elsewhere in the Universe. It could be that within the vast range of possible alien environments, all sorts of odd ecosystems and unusual but stable scenarios could arise that bring forth civilised, intelligent alien beings that look and think very differently from us. But, although these may flourish in their multitude of forms within this almost infinite Universe, the smart money is on most civilised extraterrestrials being recognisable biochemical entities with limbs, heads, sense organs and reproductive equipment not so very different from ours.

I began this chapter by asking you to imagine the Universe teeming with life. But there is of course the possibility that it does not. Perhaps we *are* alone; perhaps we are the only species to develop a civilisation, to become self-aware individuals, to crawl from the evolutionary mire and reach for the stars. Just how likely this is, is explored in the following chapter, in which I ask the question: is ours the only advanced civilisation, or are we just one of perhaps millions of societies that have developed in the Universe at one time or another?

4

WHAT ARE THE CHANCES?

'Innumerable suns exist; innumerable earths revolve about these suns . . . Living humans inhabit these worlds'

SIXTEENTH-CENTURY MONK GIORDANO BRUNO, SHORTLY
BEFORE BEING BURNED AT THE STAKE BY THE INQUISITION

The finding of what are thought to be the fossilised remains of an ancient Martian life-form on meteorite ALH84001 has raised everybody's hopes that there may be life on other worlds, but viewed dispassionately, what are the real chances of any form of intelligent life arising on planets beyond Earth?

The matter has fierce advocates on both sides of the fence, and the scientific community is clearly divided over the issue. On the one hand are those who refute vehemently the entire idea of extraterrestrial life, who claim that the evolution of life is such an incredibly complex thing requiring so many things to go right that it could not have occurred more than once in our Universe. Less extreme, but still casting a very pessimistic shadow, are those who disclaim the notion that there could be *intelligent* life elsewhere in the Universe.* Those who hold this latter view include the writer Marshall Savage, author of *The Millennial Project*, a

---

* Of course, there are those who question whether there is actually intelligent life on Earth, but that's another story.

testament outlining how humans will eventually colonise the entire galaxy, and the physicist, Frank Tipler.

But then again there are those who argue with equal passion that the Universe is so vast (to all intents and purposes, infinite) that almost anything can and does happen. They agree that the processes that produce life from inanimate matter are amazingly complex, but if it has happened once on Earth, there is every reason to believe it is possible on many of the vast number of worlds beyond our tiny solar system that we are only now beginning to discover. Advocates of this view include the astronomer, Frank Drake, creator of the first SETI (Search for Extraterrestrial Intelligence) project, and the late Carl Sagan, who believed passionately that the Universe is teeming with life and spent much of his professional career explaining the arguments to the public. Chandra Wickramasinghe, another enthusiast of extraterrestrial intelligence, goes so far as to proclaim: 'I believe there is life on every bit of the solar system that has a habitable region.'[1]

In Chapter 1 we explored the question of finding life on Mars, but before we look at a broader, galactic, or even universal scenario, we should pause perhaps to consider the matter of life existing now or at any time in the past within our own back garden – within our solar system. The key to life of an observable nature evolving anywhere in the Universe is the presence of water. Water, a very simple molecule made up of two parts hydrogen to one part oxygen, is the medium within which all biochemical processes occur. Quite simply, without water there can be no biochemical processes, and without these there can be no life according to our definition. So, first look for water and then you have a chance of finding some form of life.

And the search for water leads us to some unexpected places. Researchers from the US Air Force Phillips Laboratory have recently reported observations made using a lunar probe called Clementine which suggest there may be water-ice deposits at the bottom of certain craters on the surface of the Moon. The water arrived there originally during collisions with meteorites and comets, but because the Moon has almost no atmosphere and experiences extremes of temperature (−220°C during the lunar night and +200–300°C during the lunar day), most of the water has evaporated from these impacts and escaped into space.

The possible presence of water on the Moon has led no one to suggest that there may be life on the satellite or that there ever was the slightest chance of life evolving there, but it does offer encouragement for the future of human exploration and eventual colonisation of the Moon.

According to some astronomers, there is a strong chance that water once ran freely upon the surface of Mars. Running water may well have supported the microbes that became fossilised on ALH84001, and, say some of the researchers working on the meteorite, liquid water would have been essential in the processes that created the shapes and patterns seen around the supposed life-forms. Some optimistic astronomers even believe there is a chance water may still flow beneath the Martian surface.

Mars used to be much warmer than it is now. Today water could certainly not remain in liquid form there because the temperature drops to −123°C at the poles and −58°C at the equator during winter. However, beneath the surface, volcanic activity might just provide enough heat to produce an environment in which liquid water may still exist.

The polar caps of Mars are made almost entirely of dry ice – solid carbon dioxide, plus perhaps a tiny amount of ice water. But it would have to be a hardy organism indeed that could survive not only the extreme cold of the poles – at least twice as cold as the lowest recorded temperatures in the Antarctic – as well as exposure to a lethal cocktail of radiation unfiltered by any form of substantial atmosphere.

Other than Mars, the most likely home for life within our solar system is the Jovian moons – Europa, Callisto and Ganymede – and the Saturnian moon, Titan.

In April 1997, NASA scientists announced they had detected the presence of simple organic molecules on two Jovian moons – Callisto and Ganymede – which had been studied during close approaches of the Galileo probe. And, at the time of writing, researchers are eagerly awaiting information from a fly-by of Europa.

Europa is almost as large as our Moon, but it is very different geologically. It is approximately five times further from the Sun than the Earth, so at first it would appear to be an unlikely candidate for our search, but it harbours some interesting geophysical secrets that might radically affect the chances. Although the surface temperature on Europa rarely rises above −145°C, it is thought that, as on Mars, there may be large deposits of water far beneath the surface, where it is warmer. The evidence for this comes from the striking fact that instead of the surface exhibiting the cratered effect of our own Moon, pot-marked by billions of years of collisions, the Europan surface is incredibly smooth. The reason for this lies with the fact that the ice deposits beneath the surface rotate at a different speed to the core of the moon, and cause tensions that generate enormous seismic upheavals creating fissures miles deep. The observed smoothness on the surface of the moon is thought to be caused by liquid water bursting through miles of rock and reaching the surface through these fissures.

Another source of heat capable of melting the ice would be the presence of radioactive materials in Europa's core. A still more exotic form of heat could be generated by the close proximity of Jupiter – the largest world in our solar system. Because Jupiter is so massive, it possesses a very powerful gravitational field which has the ability to squeeze and stretch its moons, thus generating enormous amounts of heat.

Similar effects could explain how liquid water may also exist on Titan, the largest Saturnian satellite. Scientists know from spectroscopic analysis (and from NASA's Galileo mission) that water is definitely present on Titan. However, positioned as it is even further from the Sun than Europa (1.43 billion kilometres, or about eight times further from the Sun than is the Earth), some powerful volcanic effects would have to be operating to generate the required heat.

Despite these objections, for some researchers, Titan is top of the list of possible locations for life in our solar system beyond the Earth (an even better candidate than Mars); and is thought to possess at least the makings of some form of primitive biochemical systems. Steve Squyres, a researcher working at Cornell University, claims confidently that 'Titan is definitely a candidate for some fascinating prebiotic chemistry',[2] meaning that the prerequisites for some primitive form of biological activity might be present in the environment of Titan.

Whether such prerequisites have fallen into place in the right sequence and at the right time for us to observe them on Titan, Europa or even Mars, is still open to some conjecture, and will not be determined for sure until we can at least send mechanical devices to land on these worlds to make a close and thorough study.

So, as this review shows, looking at the situation optimistically, there is a chance that some form of primitive life may have existed in the past on at least one other world in our solar system. There may even still exist some form of primitive life (or at least a prebiotic state) somewhere in our neighbourhood. But, as much as it would make everyday life more exciting (and busier for exobiologists), it is perhaps best not to harbour too great a hope. A wiser approach may be to consider the broader picture – life beyond our solar system; and when we do this, the issue of extraterrestrial life and even extraterrestrial intelligence suddenly takes on an altogether different complexion.

But first, a word of restraint concerning any serious hope we may have of finding life out there in the galactic wilderness. With this discussion we are now entering a different scale of operations, a volume of space many orders larger which offers far greater scope for the search, but, as soon as we turn our thoughts and our telescopes and probes beyond the tiny confines of our solar system to the distant stars, we encounter a serious problem – the question of distance.

By our own everyday standards, our solar system is vast – around 12 billion kilometres in diameter – but even this huge distance pales into insignificance and such numbers made trivial when we begin to consider an attempt to find life on planets orbiting other stars.

The nearest star to our own sun is Proxima Centauri, which lies 4.3 light-years from Earth. This is a staggering distance. It means that light from this star travelling at just over 300,000 kilometres per second would take 4.3 *years* to get here. To convert this into units with which we are more familiar, 4.3 light-years is equivalent to a distance of:

300,000km × the number of seconds in a year × 4.3

The number of seconds in a year can be calculated by taking 3,600 (the number of seconds in an hour), multiplying it by 24 (the number of hours in a day) and multiplying this by 365 (the number of days in a year). This equals 31,536,000, a little over 31.5 million.

So, we then have: 300,000 × 31,536,000 × 4.3.

This calculation results in a little under 4×10^{13} kilometres (40,000,000,000,000, or 4 with 13 noughts after it, or 40 million million kilometres). This is roughly equal to 50 million return trips to the Moon aboard an Apollo spacecraft. Or put another way: at the speed at which the Apollo capsules travelled (about 40,000kph) it would take about 100,000 years to cover the distance to Proxima Centauri, and this is our nearest neighbour.

Our immediate problem is nothing so exotic as finding a way to get to the stars, or even to explain how an advanced civilisation could cover the distance and travel here to meet us. It is simply that such distances make it very difficult for us to know anything about other worlds. As we will see in Chapter 6, it has only been within the past two years that scientists have acquired concrete evidence that other planets orbit distant stars. At present, we know almost nothing about these worlds and have no clue as to whether they harbour life, although, based upon what we know of their proximity to their suns and their probable physical and chemical composition, many of these newly-discovered worlds appear to be unlikely candidates.

A further problem in trying to come up with any form of definitive answer, or even an approximation concerning the chances of finding life in the Universe, is that we have no clear idea of all the variables that determine the creation and evolution of life or how these inter-relate. For example: how likely is it that molecules of DNA can form given a long enough time period?

How many stars possess planets? How likely is it that even complex molecules can evolve into living material? We know all these things have happened at least once, but has it been only once, or billions of times? It is over these issues that the pro- and anti-extraterrestrial life groups are divided. The pro- camp believe that life-producing systems develop frequently, and the anti- group insist that the mechanism is so complex it could only have occurred once.

The second stage of the argument goes way beyond this, and takes on the issue of whether or not life, once started, can evolve into intelligent life-forms capable of developing civilisations and perhaps even able to communicate with one another. To try to quantify the argument, in 1961, Frank Drake, one of the pioneers in the search for extraterrestrial intelligence, produced a formula which has since become known as the Drake Equation. It is very straightforward and a surprisingly powerful tool for the astronomer, except that almost all the variables can show a range of values, and no one is yet sure what numbers to put in. It is the work of astronomers, biologists and geologists to gradually narrow down each of those numbers to something more workable and to then come up with some form of answer to the Drake Equation. The equation is:

$$N = R \times f_p \times n_e \times f_l \times f_i \times f_c \times L$$

Despite appearances, this is a surprisingly simple mathematical equation and simply requires the scientist to slot in different values for the parameters. The difficult part is working out what numbers should be put into the equation.

The letter N signifies *the number of civilisations in our galaxy trying to make contact with the human race.* The symbols on the other side of the equation represent separate factors which have to be considered in addressing the question: is there life beyond Earth? (Each term is considered in isolation; in other words, the number assigned to, say, f_p is independent of that given to L, f_i or any of the others.) When numbers for all of these factors are slotted into the equation, we end up with a value for N.

So what are these factors, and how can we assign numbers to them?

First, let's look at the term **R**. This stands for *the average rate of star formation*. A common misconception is that the Universe was formed at the time of the Big Bang, and that there has been no change ever since. But this is not the case. The prevailing theory is that the Universe is expanding, and that stars and planets are being created and destroyed constantly. Scientists are now able to observe this birth process using instruments such as the Hubble Space Telescope. It seems that some parts of the galaxy are more fertile than others, and the process of star birth is far slower than it was at distant points in our galaxy's past. Even so, making a conservative estimate, astronomers currently believe that about ten new stars are formed in our galaxy every year. So **R** is one of the variables which is pretty much agreed upon – i.e., it has a value of 10.

f_p is *the fraction of stars that are thought to be able to support planetary systems that are considered 'suitable'*. By this, astronomers mean a planetary system containing Earth-like planets. Not surprisingly, the processes involved in determining how a solar system forms is complex, and a number of factors have to be in place to produce what we are terming 'suitable' solar systems. First, the age of the star must fall into a certain range. If it is too old, its fuel will be running down and it will emit radiation that would be unhelpful for the formation and maintenance of carbon-based life. Also, as a star gets older, the rotation of planets in orbit around it begins to slow. If the star is more than around 6 billion years old (our Sun is about 5 billion years old), this will produce a dramatic effect. Planets orbiting very old stars will have stopped rotating altogether, and will have one face permanently turned towards its sun, with the other existing in permanent night. Conversely, if the star is too young, it may not have had time to allow planet formation, and the mechanism that creates and evolves life-forms will not have had time to reach a sufficient level of advancement.

However, far more important than these considerations is the question of the type of star maintaining the system. Planets that can sustain life-forms capable of developing civilisations could not be found orbiting types of stars called pulsars or quasars. These are exotic stellar objects which emit damaging radiation that would greatly reduce the chances of life forming on the planets around them. Furthermore, even if the star were a 'normal' sun, it would have to be stable over very long periods – billions of years – to allow the planets to establish their own stable ecosystems.

Finally, many stars are binary – that is, they are made up of two stars orbiting one another. Although this system by no means rules out the formation of planets, binary stars are generally considered less likely to possess systems like our Sun.

When Drake first suggested his equation, the value for f_p could only be guessed at, but recent astronomical findings have begun to narrow down the range of numbers this could be. Back in the early 1960s, Drake placed f_p at about 0.5, meaning that half the number of stars in the galaxy were potentially able to form planets.

As we will see in Chapter 6 (which looks at the latest discoveries of new planets orbiting distant stars), these findings are at once encouraging and discouraging, in that they present clear evidence of planetary systems other than our own, but they also show that the initial ideas of researchers like Drake were wildly optimistic. Rather than using 0.5 for this factor, 0.1 now seems to be nearer the mark. In other words, one in ten stars may be capable of developing and sustaining a planetary system.

Next we come to n_e, which is *the number of Earth-like planets per star*. Once again, we are bound by very limited experience. In our solar system, there is really only one Earth-like planet – the Earth. In some ways Mars comes close, and the Jovian and Saturnian moons may be viable candidates for life-supporting environments, but to be conservative, and because we still know too little about these locations, we should put n_e for our solar system at a value of 1.

f_l in the Drake formula stands for *the fraction of Earth-like planets upon which life could develop*, and in trying to assign a number for this we really are in almost totally uncharted waters. As we saw in Chapter 2, the first question we need to ask in order to arrive at a value for f_l is: what is life?

For the purposes of this book we are only interested in reaching a conclusion about intelligent life-forms with which we can communicate. It may be that any number of exotic creatures live in this almost infinite Universe, but the chances of contacting them or communicating with them is even less likely than the probability of encountering a life-form with which we can communicate. There is even the possibility that we have encountered these beings and have, for one reason or another, been totally unaware of them, or they of us. So, as I detailed in Chapters 2 and 3, to allow for 'life as we know it', we need to think in terms of carbon-based life able to communicate with us, and for a planet to be the home of carbon-based life it has to possess a certain set of environmental conditions and materials in its primeval history. Furthermore, it requires a subsequent set of finely-tuned conditions and materials for that life to have evolved and to flourish. Sceptics such as Tipler and Savage argue that these conditions are unlikely to be duplicated in the Universe, and that the probability of life evolving elsewhere is therefore slight, but as we saw in the last chapter there is a growing body of evidence to oppose this view.

According to Frank Drake, 'Where life could appear, it would appear.' He assigns a value of 1 to the parameter f_l. In other words, there is a 100 per cent chance that a suitable planet will form life if it has the correct conditions. Others, such as the Nobel Prize-winning chemist, Melvin Calvin, and the late Carl Sagan, have concurred, believing life is more likely than not to form on a suitable planet. For those who do not believe in extraterrestrial life, the value for f_l is the most crucial of all the terms in the Drake Equation. Taking the diametrically opposite view, they place its value at 0, which would consequently make N equal to 0, meaning no life anywhere else in the Universe. The value for f_l probably cannot be anything other than 1 or 0, so, for the purposes of our discussion, I will give it a value of 1.

Next we must look at f_i, the term for *the fraction of Earth-like planets where life has developed intelligence*. Again, when we first contemplate this expression, we are struck by the need to define what constitutes intelligent life.

As I mentioned in the last chapter, dolphins and whales are considered to be highly intelligent animals that could have formed a civilisation if they had evolved on land. They can communicate with members of their own species and have been known to interact cogently with humans. Attempts have even been made to decipher the complex sequence of clicks and squeaks they use to communicate with one another.

In a different sense, ants and bees act in an intelligent fashion when considered as a collection of individuals each acting as a unit in a larger society, a *Gestalt*. So, if we were to apply Drake's Equation to our planet, we could arrive at a value for f_i of between 1 and at least 4, but, again being conservative, let us take the value of 1, to represent only *Homo sapiens*.

The penultimate term in the Drake Equation is f_c. This represents *the fraction of intelligent species who would want to communicate with us*, and again, we are faced with highly speculative arguments in an effort to assign a value to our equation. In order to use this term, we have to place some limitations upon how we arrive at a value. We must first assume that an intelligent race uses some form of electromagnetic radiation with which to communicate and to interact with their Universe. Most scientists would agree that it would be unlikely for an intelligent species to have developed without using any form of electromagnetic radiation. An alien intelligence may utilise extreme regions of the spectrum – they may see in the infrared or the ultraviolet because of the nature of the light emitted by their sun. Alternatively, they may live in extreme conditions such that vision is as unimportant to them as it is to some deep-sea creatures, but whatever extreme situation might be imagined, they must use some form of electromagnetic radiation. If this is not the case, then such an alien race would fall outside the category of life as we know it.

As a civilisation, we utilise a range of radiation, from radio and television signals to X-rays, from ultrasound to microwaves, so it is likely that any civilisation at least as advanced as us would also employ similar electromagnetic waves within their technology; they may even have developed something similar to television or radio. Even if they had not created an entertainment system which leaked signals into space, as our televisions have done in recent decades, but were actively interested in communicating, they should be able to build equipment that would receive and decipher signals from space.

This then leads to a question concerning the sociological and psychological make-up of an alien intelligence. Again, this matter takes up a whole chapter later on, but for now, let us just consider the question: would they necessarily *want* to communicate with us? It is a serious point that the signals we have been sending inadvertently into space may have presented our race in a very poor light. For some seventy-five years our calling card has been television (and radio) signals conveying images of everything from the most violent Hollywood films to news coverage of war, famine and torture, leaking into space in far greater quantities than any form of contrived, politically correct message we may wish to send to our celestial neighbours. Many of these signals would be too weak to reach distant stars, but this is perhaps underestimating the sensitivity of possible alien detection systems. Television and radio signals are no different from any other forms of electromagnetic radiation in that they travel at the speed of light. It is therefore conceivable that alien civilisations living on planets up to seventy-five light-years away could be chuckling at our antics, or perhaps battening down the hatches for fear we would lower the tone of the neighbourhood. Furthermore, because their reply would take as long to get to us as our signal took to reach them (nothing can travel faster than light), if we received a message tomorrow from a civilisation twenty-five light-years away, they would be responding to signals from our world that were fifty years old – footage of the D-Day landings, perhaps.

So, what value do we give f_c? On the one hand, it would seem likely that any civilisation would eventually develop a form of long-distance receiving and transmitting system using electromagnetic radiation and enabling communication, but how many races would want to make contact? There could be an abundance of species busily communicating with one another but excluding us; equally, alien civilisations could prefer to keep themselves to themselves whether or not they have been warned off. Weighing up these factors, a conservative estimate would be that 10–20 per cent of intelligent aliens would want to communicate, so f_c would be, say, 0.1.

Finally, we come to the last term in the equation – L. This represents *the lifetime of a civilisation* (in years). And again, in attempting to assign a value, we face another complex series of permutations. To assign a value to L, we have the difficulty of having to consider hypothetical sociological factors for a hypothetical race, but we have again one example to draw upon – our own experience.

It suggests an interesting synchronicity that our race developed weapons of mass destruction at almost the exact point we revealed ourselves to the Universe with our electromagnetic signals, and it could be that many races are destroyed about the time they could make contact with their neighbours. Since Frank Drake first devised his formula in 1961, there has been much debate among scientists about the value of L, and during the past four decades the political and social zeitgeist has altered radically. The Cold War has ended, but the threat of nuclear destruction is still very much with us and the killer instinct of humankind has not changed in the slightest. Perhaps the ease with which human beings make war is irreducibly linked with our drive to progress and advance. It is possible the instinct that drives us to communicate derives from the same source as our aggression. If this is the case, it may be the same everywhere, and it would be reasonable to assume a large proportion of civilisations destroy themselves at around the time they develop the technology to communicate with beings beyond their own world.

But war is not the only means by which a civilisation can meet its end. Scientists are only now beginning to realise the very real danger of planetary collisions with comets or asteroids. It is believed that a devastating asteroid collision caused a sudden traumatic alteration in the ecosystem of the Earth some 65 million years ago, resulting in the extinction of the dinosaurs, and there have even been a number of documented near-Earth collisions this century. The massive explosion reported in Siberia in 1908, which devastated hundreds of square kilometres of forest in the region of

Tunguska, is thought to have been caused by a meteorite explod-
ing several kilometres above the ground. If this had occurred over
Paris or London, millions would have died. An object only a few
times larger than the Tunguska fireball impacting with the Earth
would not only devastate a wider area, but the dust thrown up by
the collision could produce a blanket around the entire planet
capable of destroying all life on the surface. If it landed in the sea,
which is more likely, the tidal effects could be even more
damaging.

There is also the question of planetary resources. As a race, we
are perilously close to over-exploiting the resources of our world,
and we are already capable of severely damaging planetary mech-
anisms that are there to maintain an ecological balance. It is
conceivable that other civilisations have followed the same path
and gone further, completely destroying their own environments.
Such threats as reduced fertility, AIDS, superbugs and nuclear ter-
rorism are all further potential civilisation-killers, and I'm sure
there are plenty more around the corner of which we are still
blissfully ignorant.

One theory concerning the longevity of civilisations suggests
that they either survive little more than 1–2,000 years or else they
continue for perhaps hundreds of millennia. It is possible that
many races pass through a 'danger zone', during which they have
a good chance of destroying themselves, but if they come through
it, they develop into highly advanced cultures capable of inter-
stellar travel and galactic colonisation.

In an extreme case, L also depends upon astronomical factors.
If we assume that life may form on a large number of planets, and
that those life-forms could evolve into intelligent civilised beings,
the time at which life began on their world would be a crucial
consideration.

Our Universe is believed to be approximately 12 billion years
old, and our Sun is a very typical star located some two-thirds
of the way along one of the spiral arms of the Milky Way
galaxy – itself an 'ordinary' galaxy among an estimated 100 bil-
lion others. In astronomical and geological terms, the Earth is
quite average, and life began to appear here a little under 4

billion years ago, or around 8 billion years after the Big Bang. But it is quite conceivable that a great many planets around other, older stars cooled long before our own planet. Astronomers have observed the death of stars far more ancient than our own. If any planet around these stars had brought forth life, a civilisation that formed there would either be ancient, interstellar voyagers or long dead.

To find a sensible value for L, we must assume a normal distribution of ages for successful civilisations. If L for a particular planet is 2,000 (years), the race may have destroyed itself and would therefore be of no further interest to our argument. But L could be much larger. It is possible there have been and still are civilisations hundreds of millions of years old. Equally, there could be a large number of very young civilisations, perhaps no more than 2–3,000 years old. Most civilisations that have survived and are able to communicate would be somewhere between the two extremes.

Drake and his colleagues have placed a value of 100,000 on L. This seems rather arbitrary, but nevertheless, if we put any figure over the 2,000-year watershed, the equation still gives us a correspondingly large number for N – which, remember, is the number of advanced civilisations wanting to make contact. So, let's start assigning numbers:

R has a value of 10
f_p = 0.1
n_e = 1
f_l = 1
f_i = 1
f_c = 0.1
L = 'a large number' (ranging from 2,000 to perhaps many millions)

Putting these figures into the Drake Equation, we arrive at an interesting result:

$$N = 10 \times 0.1 \times 1 \times 1 \times 1 \times 0.1 \times \text{[a large number]}$$

The 10 and the 0.1 cancel each other out, giving us $1 \times 1 \times 1 \times 0.1 \times$ [a large number]. Which is then equal to: $0.1 \times$ [a large number]. Which is still a large number! But how large?

If we call L (the average age of a civilisation) 100,000, it means there are 10,000 civilisations sharing just this single galaxy (one of 100 billion, remember).

Frank Drake believes the value of L to be much greater than 100,000, which would mean that N would be correspondingly larger. According to some enthusiasts, N could be in the tens or even hundreds of millions.

This may seem excessive, but when we consider that our galaxy contains upwards of 400 billion stars (that is, 400 thousand million), then an estimate of 100 million civilisations means that there is only one such race for every 4,000 stars.

In some parts of the galaxy, where the stars are more densely packed than out here on the edge of the Milky Way, 4,000 stars may be packed into a sphere no more than a couple of light-years in diameter. In which case communication might be easier. Even in our region of the galaxy, a sphere fifty light-years across contains many stars, some of which are similar to our Sun. Which leads us to the inevitable question: if so many planets are home to advanced civilisations, why have they not made contact with us?

5

SIGNALS FROM
BEYOND

'It's good to talk'

BT ADVERTISING SLOGAN

In one of the most impressive openings shots in modern cinematic history, long shadows fall across the American flag planted by the Apollo astronauts on the Moon and an alien craft moves towards Earth. Seconds later, alarm bells ring in the control room of a giant detector we are told is part of the Search for Extraterrestrial Intelligence project. And so *Independence Day* begins, taking us on a rollercoaster, multimillion-dollar B-movie ride into the jaws of global conquest and beyond.

However, this is almost certainly not how we will first make contact with an alien civilisation.

The earliest evidence for the presence of another intelligent race alive and well in the Universe will probably come from a SETI detector somewhere in the world, but it will be a signal from a distant star that will alert us, not the electromagnetic presence of a gigantic spacecraft passing inside the orbit of the Moon. Just to prove their point, spokespeople from SETI have their own website, called 'SETI and *Independence Day*', in which they explain the difference between them and the fictional version, and, with tongues half in cheek, invite the makers of the film to contribute a tiny percentage of *Independence Day*'s profits to the on-going (and very expensive) search for extraterrestrial intelligence.

The idea of searching for life beyond Earth has been around for a long time. In the Roman era the poet and philosopher Lucretius wrote: 'Why then, confess you must, that other worlds exist in other regions of the sky, and different tribes of men [and] kinds of wild beasts.' In AD 165, the historian and dramatist Lucian wrote about the possibility of a lunar king and queen whose people were at war over the colonisation of Jupiter; and the Italian monk Giordano Bruno was burned at the stake in 1600 for declaring a belief that other planets existed beyond the Earth, and that these might be inhabited by beings not so different from ourselves. It is estimated that between Lucian and the beginning of this century, almost two hundred learned books were written on the subject of extraterrestrial life, a number which can be multiplied by several factors of ten during the twentieth century.

Naturally, none of the writers and philosophers who were deeply moved by the idea that life could exist beyond our world could do anything about finding any form of alien presence in the Universe. Indeed, it was not until a couple of years after the launch of Sputnik I, after mankind had entered into its own Space Age, that the earliest primitive attempts were made to detect signals from an alien civilisation. And, like many innovative scientific ideas, the concept of building detectors to pick up signals from another civilisation must have been 'in the air' during the spring of 1959, because at least two totally independent teams were thinking along the same lines at that time.

At Cornell University in the United States, two physicists, Giuseppi Cocconi and Philip Morrison, were composing a paper that was soon published in *Nature* in which they suggested that microwave radiation could be used as a means of communicating between civilisations. Meanwhile, at a radio telescope in West Virginia, a young astronomer named Frank Drake was attempting to persuade his superiors to let him have some telescope time for an ambitious attempt to detect radio signals coming from possible worlds out in the depths of the galaxy.

The principle behind SETI is that any civilisation advanced

enough to have invented some form of radio or television would at least be leaking radiation from their home world in the same way that we do. Alternatively, they could be sending signals deliberately in order to see if anyone was out there. After all, we are so preoccupied with the idea that we might be the only civilisation in the Universe, it is easy to forget that there may be thousands or even millions of cultures out there where individuals are wondering if they too are alone.

It is fairly safe to assume that alien intelligences would be using radiation in much the same way we do. The Universe is awash with radiation running the full range of the spectrum. It is very likely that alien beings would see and hear in a different range of the spectrum from us because their home sun might well be emitting a slightly different mix of radiation and their ears may be quite different (although, once again, the radiation could not be extreme since this would certainly be harmful to carbon-based life). If such a culture has emerged on a world where radiation is freely present and used as part of the ecosystem, then a technologically advanced civilisation would probably go through a 'Radio Age' and 'Television Age' or something not totally dissimilar, and, if they were interested in contacting others, they too would set up transmitters and receivers.*

Naturally, these signals would be weak by the time they arrived on Earth. Indeed, recent research has suggested that gas clouds between stars may be more dense than previously believed, and that the power of alien signals could be weakened, perhaps preventing them ever reaching here.

Like all matter and energy in this Universe, the behaviour of these signals would be governed by relativity, and restricted to the same speed as all electromagnetic radiation – the speed of light

---

* There are many ways in which nature utilises radiation. Photosynthesis is one obvious example, where plants use specific frequencies of radiation to initiate reactions to produce energy. On Earth, most creatures use radiation to see, and they hear using sound waves – another example of how information is transmitted by 'radiating' or via a wave-form similar to radiation within the electromagnetic spectrum. Even such exotic creatures as bats, which cannot 'see' in the way we do, use radiation to operate their 'radar'.

(300,000 kilometres or 186,000 miles per second). So, any signals we might one day receive will have been travelling here for many years, and may even be coming from a civilisation so far away that the race which produced them has since become extinct or totally transformed.

Working on little more than a hunch, Cocconi and Morrison at Cornell and Frank Drake decided that the best approach for their studies would be to search within a narrow frequency range bordered by one particular frequency – 1420 MHz – and to home in on just a few carefully chosen target stars.

This second decision was really dictated by the very limited nature of their equipment. In those days, at the start of what would become the SETI project, the effort was little more than a hobby, an activity totally dismissed or ignored by the establishment and conducted outside official working hours with whatever resources were available. In 1960, when Drake first received official sanction to devote some of his time to SETI research, the equipment available could not have allowed him to do anything other than to concentrate on a couple of targets. But why choose a frequency of 1420 MHz?

Known as the 'waterhole', the chosen frequency range centred on 1420 MHz is a quiet, 'dark' part of the spectrum but bounded by two very important universal constants. Naturally-occurring free hydrogen is known to transmit a signal at precisely 1420 MHz, and close by in the spectrum (at 1670 MHz), a molecule called hydroxyl (OH^-) also transmits a natural signal.* To the chemist, H and OH are a magical combination – together they make H_2O – hence the name 'waterhole' for the gap in the spectrum between the frequencies of the components that go to make water. But the decision was not due simply to poetic licence. Because hydrogen is abundant throughout the Universe, both Drake and the pair from Cornell assumed that any alien intelligence advanced enough to have developed radio technology would know of this chemical fact and expect a recipient to be

---

* Strictly speaking, OH is not a stable molecule. On Earth it only exists as OH^- and is known as a hydroxyl ion.

equally knowledgeable – it was seen as a code, an indication of 'intelligence', or an awareness of a basic, communicable scientific principle.

So, early in 1960, with unalloyed enthusiasm, a budget of $2,000, and very little moral support, Frank Drake set up the earliest practical SETI, an effort he named 'Project Ozma', and began hunting for signals in the 1420–1670 MHz region.

The biggest difficulty Drake and other researchers faced in the early days, and one that still confronts researchers today, was the problem of interference. Because SETI scientists are trying to detect what must inevitably be very weak signals sent from points great distances from Earth, the presence of electromagnetic radiation on their own doorstep or from the very machines they are using has to be carefully monitored and taken into account. In the early years there were numerous claims of contact, but after the initial excitement died down these were revealed as electromagnetic leaks from equipment or outside 'local' interference. In one memorable case, a low-flying aircraft passing directly overhead at irregular intervals produced a very strong signal that appeared to come initially from a distant star named Tau Ceti.

Such enthusiastic amateurism defined SETI throughout the 1960s and 1970s, but a growing number of respected scientists and researchers from disparate fields became interested and increasingly vocal in their attempts to get decent funding with which to establish a co-ordinated SETI project. Although this was to prove to be some way off, several useful advances were made during the 1970s and early 1980s.

At NASA's Ames Research Centre in Mountain View, California, a collection of enthusiasts developed a project called 'Cyclops', which looked principally at the theoretical aspects of contacting extraterrestrial intelligence – how to detect alien signals and where best to look. They then went on to develop better, dedicated equipment. But still the budget was tiny, and the enthusiasm was never matched with hard cash.

During the 1960s and 1970s the Russians made rapid progress with their own SETI research and were, in many ways, far ahead of their Western colleagues. Scientists, uninterested in the political wrangling of the era, somehow managed against the odds to organise international conventions and to share research material as best they could, but until recently there was no international or concerted effort.

Our best hope of contacting alien intelligence lies with the latest incarnation of the SETI project. This is called 'Project Phoenix', and it gives researchers the first chance to set up detectors around the world at different sites and to use any 'spare' time available at the major radio telescopes and observatories of the world. Phoenix was originally funded by the US government, who had finally been convinced by the arguments of such luminaries as Carl Sagan that a serious, well-funded SETI project would cost almost nothing compared to the annual NASA budget. They argued that it was not merely a scatter-brained waste of public resources used for the personal satisfaction of a few s.f.-influenced scientists, but should be considered a viable research effort that could bear fruit in allied fields.

The plan was to create a $10 million-per-year programme. This began in 1992, but little more than a year after it was established, funding was terminated by the interference of Congressmen who had changed their minds again and had decided that they were paying for the indulgence of UFO cranks. So, Phoenix languished for another year until funds were raised from philanthropic enthusiasts and public interest. Rather aptly, Steven Spielberg, director of *ET: The Extra-Terrestrial* and *Close Encounters of the Third Kind*, is funding a project based on the east coast of the United States which is linked with Phoenix, and other wealthy enthusiasts are putting money into the search at a variety of sites around the world.

Project Phoenix is a pan-global operation using the world's largest radio telescopes to sweep the Universe between 1000 and 10,000 MHz (1–10 GHz) – which happens to be the region where there are many natural resonances and emissions (including, as we have seen, that for hydrogen). NASA have also developed what they call spectrum analysers to search across a wide range of frequencies within the chosen limits, and are currently devising software to filter out noise and other interference in the signal.

The development of suitable technology is one great hurdle for
the detection of interstellar signals, but there are other major
questions to be answered when setting up such a scheme. SETI
research is like no other form of scientific study. In almost all
other disciplines, scientists have something concrete upon which to
establish their theories and to develop plans. Even within the
world of quantum mechanics, physicists have the resources of
particle accelerators, such as those at CERN near Geneva or
Fermilab outside Chicago. They also have the rigorous framework
of mathematics as a guide. The SETI researcher is looking for a
needle in a very large haystack, a needle which may not be there
at all, and they are trying to find that needle with the flickering
light from a match on a windy night while wearing sunglasses! But
as the original thinkers in the field, Cocconi and Morrison, said in
their seminal paper back in 1959: 'The probability of success is dif-
ficult to estimate, but if we never search, the chance of success is
zero.'[1]

This is undoubtedly true, but the technical difficulties are not
the only problem. The first dilemma is: where should the receivers
be pointed, and should they be aimed at a few well-chosen stars
or should they be used to sweep the heavens?

Project Phoenix is attempting to solve this problem with what
they hope will be as close to a 'catch-all' approach as funding will
permit. They are planning to conduct three different searches
simultaneously, involving 1,000 selected stars of various types.

The first of these searches is called the 'Nearest 100 Sample'.
As the name implies, this sample includes the hundred stars near-
est to our Sun, all of which lie within twenty-five light-years of
Earth. The downside of this selection is that many of them are
thought to be less likely to have life-bearing planets orbiting
them than some other, more distant suns. On the plus side, if any
of these suns possess life-supporting planets, then in interstellar
terms they are within our own back garden and their proximity
would make contact and communication a little easier. This
group of hundred stars includes some familiar names – the near-
est stars to Earth – Proxima Centauri and Alpha Centauri (a
binary star system); some suns that are a great deal larger and

hotter than ours, such as Sirius; and at least twenty-five other binary systems.

The second category is 'The Best and the Brightest Sample', involving 140 stars all within sixty-five light-years of Earth. In some respects this group is seen as perhaps the most likely to give positive results, and includes many of the stars recently proven to have their own planetary systems. Other stars in this group have been selected because they are thought to be within the age range most conducive to supporting planets that may be home to surviving civilisations. This sample has also been narrowed down to focus on a large proportion of single stars, as opposed to multiples, where two or more stars are in close proximity – a situation that produces problems for the chances of life developing on planets orbiting them.

The brightness of a star can be misleading. It is natural to think that all bright stars are very close to us, but in many cases this is not so. No two stars are the same; they vary enormously in their chemical nature, size and brightness. Sirius, for example, the brightest star in the sky as seen from Earth, is actually many times further away than a relatively faint star such as Proxima Centauri, our nearest neighbour. In fact, despite being so close, Proxima Centauri emits so little light that it is one hundredth the brightness of the faintest star we can see without a telescope. 'Radio brightness', which is what SETI researchers are primarily interested in, follows the same pattern – it does not necessarily equate to nearness.

The third selection of stars, called the 'G-Dwarf Sample' is the largest group, and is a wider range of stars, but they all lie within 200 light-years of Earth and are what astronomers call G-Dwarfs, a category into which our own Sun is placed. All the stars in this group are thought to be very like our Sun in terms of age and size. They are almost all single stars and constitute the largest grouping of all those being studied by the Phoenix researchers – some 760 stars.

Phoenix's observing system is designed to be transported around the world and linked up to any available radio telescope powerful enough for the purpose. The first venue for Phoenix was

Australia, where astronomers used the Parkes 64-metre antenna and the Mopra 22-metre antenna, both in New South Wales. Because Australia was the first site, a very high proportion of the stars in the targeted group were those seen only in the Southern Hemisphere, including 650 G-Dwarf stars. In 1996, the system was taken back to the National Radio Astronomy Observatory in West Virginia, where a 40-metre dish was used to follow through the next stage of the search. The project is currently established at the largest radio telescope in the world – the 305-metre Arcibo radio telescope in Puerto Rico.

At the time of going to press, the interstellar 'airwaves' remain silent, but no one involved in the Phoenix project thought there would be much chance of immediate success. And indeed, there are some astronomers who suggest that the official SETI teams are going about things the wrong way. They argue that radio telescopes should be turned towards the centre of the Milky Way, where the stars are far more densely packed and where, they say, there is a far greater chance of finding something interesting. But this has associated problems, not least of which is the fact that it would be very difficult to separate the multitude of natural signals constantly emitted from so many stellar objects. As the British astronomer Michael Rowan-Robinson says: 'Looking along the plane of the galaxy, like looking at car headlights in a traffic jam, makes it very difficult to detect one source of radio emission from another. And, if such radio emissions would also fade away over distance, we would probably detect nothing.'

An alternative argument is that we should not be looking for radio signals at all. Some researchers suggest that an advanced alien race would have dispensed with radio long ago, and may be sending information using lasers. Others assume that the majority of surviving civilisations in the Universe would be far in advance of us and might be located by searching for the heat they generate as a by-product of their energy-production systems.

The eminent American physicist, and one-time associate of Albert Einstein, Freeman Dyson, who works at the Institute of Advanced Study in Princeton, has proposed a scheme by which a very advanced technology could produce an almost limitless fuel supply. He speculates that a sufficiently developed civilisation could harness the total energy output of their home sun by building a sphere of receivers and energy converters around it. These 'Dyson spheres', as they have become known, would of course provide tremendous amounts of energy but would also radiate commensurate amounts of heat, which could be detected light-years away in the infrared region of the spectrum. Others have taken this idea even further by suggesting that civilisations perhaps millions of years in advance of our own could utilise the energy output of an entire galaxy, or even a cluster of galaxies, and that some of the many types of energy source we see in distant parts of the Universe are the waste products from such processes.* This has led those involved with SETI to categorise potential civilisations into three distinct classes.

Type-I cultures (which include us) are those which have developed to the point where they can exploit the natural resources of a single, home world. A Type-II civilisation would be capable of building something like Dyson spheres and processing the entire energy output of their sun. This level of development would almost certainly be associated with the ability to travel interstellar distances. Such cultures may also have developed means by which they could circumnavigate the hurdles presented by the light-speed restriction. A culture that had reached this stage of development would be thousands or perhaps tens of thousands of years in advance of us.

A Type-III civilisation would be millions of years ahead of us,

---

* For more than twenty-five years, astronomers have been observing sudden bursts of energy from a variety of different locations in the cosmos. They detect these bursts, which are thought to be the result of the most powerful explosions ever witnessed, by following a left-over trace of gamma rays (a form of electromagnetic radiation) that reach the Earth. There are literally hundreds of theories that attempt to explain these bursts, including the notion that they could be the result of the activities of some super-civilisation. Recently, one such burst was carefully monitored and found to have come from an explosion so powerful that in ten minutes the source produced more energy than the total output of our Sun during its lifetime. Astronomers are actively chasing the source and the cause of this phenomenon and hope to solve the mystery after one more sustained observation of the effect. The trouble is, no one knows when or where the next one will be.

and would have developed the technology to utilise the entire resources of their galaxy, an ability which to us appears God-like but is actually possible within the laws of physics. It is nothing more supernatural than a consequence of a life-form starting their evolutionary development a little before us in relative, universal terms. To us, such beings would demonstrate God-like powers, but they too would have originated in a slurry of single-celled organisms on some far-distant planet. They would simply have had a longer time in which to develop.

This classification was first postulated in the 1960s, quickly becoming an internationally accepted standard. This was also the most active period of Soviet work on the search for alien civilisations, and on one occasion scientists in the USSR actually thought for a while that they had encountered a Type-III civilisation.

It was 1965, the Russians were leading the world in efforts to detect messages from ETs, and their top researcher was a man named Nikolai Kardashev (who was also the first to discuss seriously the idea of super-civilisations and civilisation types). One morning at the Crimea Deep Space Station, Kardashev's team detected an incredibly strong signal that was certainly of extraterrestrial origin. The interesting thing about it was not simply its power, but the fact that the signal seemed to slowly change frequency over time, sweeping through a broad band. This type of signal was quite unprecedented, and to the Soviet team almost certainly the fingerprint of a civilisation attempting to make contact.

Against his better judgement, but bowing to pressure from his colleagues, Kardashev decided to announce the finding publicly, declaring to the world's press that the source was almost certainly an extraterrestrial civilisation. Sadly, it was not to be. Within hours, scientists at Caltech in the US contacted their Russian colleagues to inform them that what they had observed fitted exactly the description of an object they too had detected a few months earlier and had been studying ever since. They called the source a 'quasar', or quasi-stellar object, and it was definitely not a signal from an advanced civilisation of any description.

Quasars are still only partially understood. Scientists know that they are tremendously powerful sources of electromagnetic radiation and that they are moving away from us at high speeds. They are believed to be extremely turbulent galaxies – a seething mass of matter and energy very different from our own stable Milky Way. It is suspected that at the heart of each quasar lies a black hole which traps within its intense gravitational field anything that approaches it. As matter and energy are sucked in, but before they disappear behind what physicists call the 'event horizon' (from which there is no return), they collide with other forms of matter already trapped there and emit energy that may just escape the gravitational clutches of the nearby black hole.

Quasars are fascinating and exotic stellar objects, and their close study has provided new insights into the nature of the Universe; but they are not the only strange objects to be discovered by accident and mistaken for the hallmarks of extraterrestrial intelligence.

In 1967, a Ph.D. student at Cambridge University named Jocelyn Bell detected a strong, regular signal coming from deep space in the waterhole region of the spectrum. After reporting the findings to her supervisor, Anthony Hewish, they agreed they would not go public until they had investigated the signal fully. Gradually they eliminated all possible conventional sources until they realised that the signal was actually an emission from a strange object in deep space that was sending out an almost perfectly regular pulse. The object was then found to be a neutron star, or 'pulsar', the remains of a dead star that had collapsed under its own gravitational field so much that the electrons orbiting the nucleus of the atoms making up the star had been jammed into the nuclei and fused with protons to form neutrons. This super-dense matter emits pulses with such regularity that pulsars are thought to be the most accurate clocks in the Universe.

Since Bell and Hewish's discovery, other regular signals have been detected which have not originated from pulsars or any terrestrial source, but have appeared only once. A team led by Professor Michael Horowitz at Harvard University has reported thirty-seven such signals during the past ten years, all within twenty-five light-years of Earth, but because they have not been repeated they do not qualify as genuine candidates for signals from a race trying to contact us. They could, of course, be one-off leakages from specific events, but we might never know, and for scientists to analyse a signal properly, they need a repeated, strong, regular pulse.

So far, the most important find was a signal detected at the Ohio State University 'Big Ear' radio telescope in August 1977. Known by SETI researchers and enthusiasts as the 'Wow' signal, after the monosyllabic exclamation written on the computer print-out by an astonished astronomer at the station, it lasted exactly thirty-seven seconds and appears to have come from the direction of Sagittarius. Although, most strikingly, the signal was a narrow-band signal precisely at the hydrogen frequency of 1420 MHz, it has not been detected even a second time, in Sagittarius or anywhere else.

So, what of the future? Is the continuing search for intelligent life in the Universe a total waste of money, as its opponents insist, or are we perhaps on the threshold of a great discovery?

In commercial terms, SETI is potentially the greatest scientific bargain ever. The cost of the project to the US government was a tenth of 1 per cent of NASA's annual budget and is now financed privately, so even the die-hard sceptics cannot claim that it is drain on the tax-payer. Furthermore, the potential gains from the success of the project would be unparalleled in human history. Quite simply, there is absolutely nothing to lose in trying.

More problematic will be maintaining the momentum of a project which, year after year, fails to deliver the goods. The argument against this is that both pulsars and quasars were discovered indirectly through the efforts of SETI researchers, and it is also true that improvements in techniques and development of new types of equipment used in the search will filter down into other areas of research and then on to everyday use.

However, one difficulty for future researchers will be the growing level of terrestrial interference. Some enthusiasts argue that we are currently living through a window of opportunity in the search for extraterrestrial intelligence, and that the embryonic communications revolution will soon work against our chances of detecting a pure signal from another world.

A solution to this would be to develop sophisticated detection systems placed outside the Earth's atmosphere. SETI and NASA have tentative plans for orbiting detectors and even devices that might be placed at the outer limits of the solar system, away from all forms of interference generated by technologies on Earth. Naturally, these are little more than dreams because of the huge cost involved. If NASA cannot find a few million dollars annually to support a terrestrial-based search, they are hardly going to finance sending detectors into deep space.

There are a few ways around this dilemma. One hope is that interplanetary unmanned travel will become very much cheaper in the near future. Thanks to the success of the shuttle, the costs of near-Earth projects have been slashed and NASA are currently developing new, cheaper systems to deliver probes to the Moon and the inner planets. Another possibility is that SETI may be able to hitch a ride with the official projects. In a similar way to Frank Drake's first efforts and his $2,000 budget back at the start of the Space Age, it might be possible to use the data collected by official researchers to search for anything untoward that may indicate the presence of an extraterrestrial civilisation trying to make contact.

At the other end of the spectrum, the demise of the government-funded SETI project has spurred on a growing group of spirited and capable amateurs. The SETI League, based in New Jersey, are establishing a network of individuals around the globe who have their own detectors and processing equipment.* The League now has twenty-seven dishes pointing to the heavens. They hope this number will grow exponentially within the next few years, and as costs come down, they expect to have 1,000 dishes by the end of the millennium.

Unlike the professionals who employ the world's largest radio telescopes, the members of the SETI League use domestic satellite dishes little different from those used to pick up television pictures. The big dishes can only be pointed at a collection of specific, carefully-chosen locations, whereas the smaller domestic dishes are able to sweep the sky. In this way, the SETI League

---

* The League may be found on the Internet at *http://seti1.setileague.org/homepg.htm*. It will help anyone interested in setting up their own detection station and will provide free software as required.

hope to generate a 'critical mass of dishes' – at least 1,000 – so they can divide up the sky into individual regions, with each area monitored by a single dish. This they see as working in perfect harmony with the big boys, who use the multimillion-dollar receivers to focus on a selected catalogue of prime targets.

Finally, we have to consider the possibility that ETs will notice us before we notice them. One way in which this may happen is that another civilisation may pick up one of our own probes.

There are many difficulties associated with interstellar travel (see Chapter 8), but we should keep in mind that the human race has actually just entered the age of interstellar travel, albeit in an ersatz fashion. Pioneer 10 left our solar system in 1997 and will continue on its journey, out of radio contact with Earth-bound stations, until it meets whatever end awaits it. It will be preserved by the fact that it is travelling through a near vacuum, but it may become severely damaged by micro-meteorite collisions or the gradual erosion of high-speed particles that smash into it from

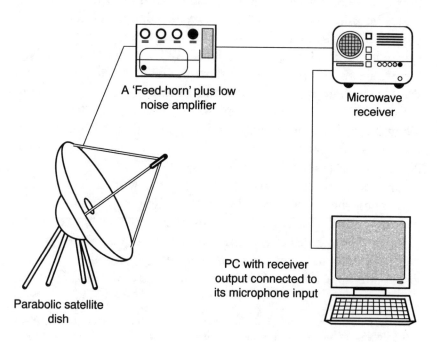

A 'Feed-horn' plus low noise amplifier

Microwave receiver

Parabolic satellite dish

PC with receiver output connected to its microphone input

Fig. 9: An amateur SETI system.

time to time. But it is possible that perhaps hundreds of thousands of years in the future it will become ensnared in the gravitational field of another star and drawn into orbit. This sun may possess planets that are home to intelligent beings who may detect the device. If this happens, they will soon know a little about us. On the side of Pioneer 10 there is a plaque (put there at the request of Carl Sagan) which identifies our world, its location in the galaxy and the appearance of humans.*

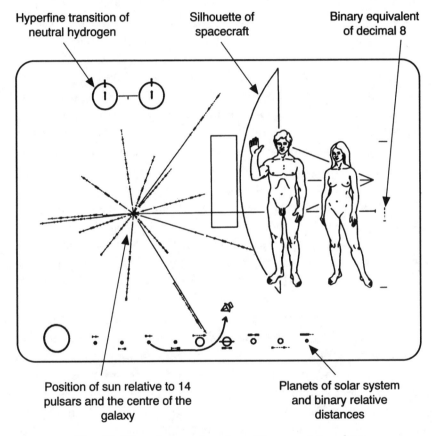

Fig. 10: The plaque placed on Pioneers 10 and 11.

---

* It may be that by this time the human race will have ceased to exist, or be star-travellers ourselves. It is even possible that we will re-encounter our primitive satellite at some point in the future, when we overtake it on its steady way to some distant star.

Pioneers 10 and 11 were followed into space by Voyagers 1 and 2, which will also leave the solar system in the near future and head off into interstellar space. These craft contain further information about the Earth and the human race in the form of a gold-coated phonograph record. The record contains 117 images taken from around the world to describe the planet and the fauna and flora found here, greetings in 54 different languages, and a representative selection of 'the sounds of Earth', including a 90-minute selection of the world's music (Mozart and Bach for the classically-inclined aliens and Chuck Berry for ETs who like to boogie).

In the final chapter of this book I will look in detail at what contact could mean for the human race, but it is perhaps sobering to realise that we have actually been a beacon for those looking for life for the past 350 million years. It was at this point in the history of our world that any advanced civilisation with technological skills equivalent to ours today would have been able to detect the presence of high concentrations of oxygen in our atmosphere. Using spectroscopic analysis they would also have been aware of our ozone layer.

For some, this offers further evidence for the view that we are entirely alone. If other civilisations, many of whom would have begun developing earlier or faster than us, have been observing the Universe for some time, then why, say the sceptics, have they not spotted us? I have tackled this argument already and will return to it in Chapter 10, but for the moment, we should turn our attention to another search that has been gaining momentum in recent years and may overtake the radio astronomers in their quest to detect signals from other worlds. As the SETI researchers have pointed their devices to target stars and the amateurs have begun setting up their domestic detection systems, other astronomers have begun to find a growing collection of planets orbiting distant stars. And where there are planets, there may be life.

6

THE GREAT PLANET HUNT

'Why, sometimes I've believed as many as six
impossible things before breakfast'

LEWIS CARROLL, *THROUGH THE LOOKING GLASS*

Look up to the heavens and at different times of the year on a clear night you might see three, maybe four planets with the naked eye. Using a small telescope you could perhaps add a couple more to the list, but it has only been since the discovery of Pluto in 1930 that humans have managed to create what we think is a complete picture of our planetary system, including all nine planets and an asteroid belt. Even today astronomers are still adding newly-discovered moons and asteroids. There should be little surprise then that as recently as 1995 scientists still had no conclusive evidence that there are planets orbiting other far-distant suns, suns many millions of times further away from us than Pluto, our most distant relative in the solar system. If evidence to prove the presence of extraterrestrial life remains elusive, this breakthrough has shown that our solar system at least is definitely not unique.

The search for planets beyond our solar system began over fifty years ago (not long after Pluto was discovered), but because of the huge distances involved, it was almost impossible to find clear evidence with the equipment then available. In 1983, a clue to the existence of extrasolar worlds came in the form of unexpected radiation emitted by a group of stars studied by IRAS (the 'infrared astronomical satellite'). Astronomers came to the conclusion that this irregularity was caused by dust clouds around the observed stars. But because such dust clouds indicate planet formation, even this relatively modest discovery was sufficiently exciting to spark the interest of several teams to probe further. And new dust clouds are now being found regularly by teams using the powerful Hubble telescope, which since its installation has imaged several nearby stars and been employed to observe dust clouds surrounding these suns.

Soon after these dust clouds were first seen, there was a further boost to the hopes of the planet-hunters when American astronomer Aleksander Wolszczan, who was working at the huge Arecibo radio telescope in Puerto Rico, detected an unusual set of objects in the heavens. At the time he was studying pulsars, which, as we saw in the last chapter, are super-dense neutron stars – the stellar debris left behind after a supernova. Neutron stars spin rapidly, emitting a pulse of radio waves like a lighthouse sending out rays into the void. Wolszczan noticed that one of the stellar objects in his study, PSR 1257+12, was pulsing irregularly. After further investigation, including a calculation of the degree of variation in the pulse, he reached the conclusion that there are two, perhaps three, planets orbiting the neutron star. The two he is sure of are about three times heavier than Earth; the possible third planet may be a smaller one about the size of our Moon orbiting close to the star.

Although this was in itself a highly significant discovery, it offers little encouragement to those searching for extraterrestrial life. The planets orbiting PSR 1257+12 could only have formed *after* the supernova for the simple reason that supernova explosions would reduce all matter close by to a soup of fundamental particles. So, even if we ignore the intense radiation emitted by pulsars and the unstable nature of the neighbouring environment, the planets Wolszczan had detected would be too young to support life.

But this work laid the foundations for more significant finds.

The truly ground-breaking discovery in this field – the detection of a planet orbiting a conventional star – came in 1995, when Michel Mayor and student Didier Queloz, working at the Geneva Observatory in Switzerland, announced the results of their search. This was the culmination of research into a small group of stars that are, in cosmological terms, relatively close to our solar system.

Like many great discoveries in science, the first extrasolar planet was noticed almost by accident. Mayor and Queloz were not actively looking for new planets, they were studying strange stellar objects called 'brown dwarfs'.

A brown dwarf is a failed star. Stars form from a cloud of gas and dust which coalesces to generate a self-sustaining system. Suns generate heat by nuclear fusion, but for the fusion process to continue producing energy over billions of years, the stellar object has to be above a minimum size and density. Brown dwarfs may be thought of as intermediates between stars and planets, or simply as stars that do not shine. Although it is difficult sometimes to differentiate between some stars, some brown dwarfs and some planets, a rule of thumb is that all brown dwarfs so far observed are at least thirty times the size of Jupiter (the largest planet in our solar system but still only one thousandth the size of our Sun). They are also entirely gaseous (like stars), whereas planets such as Jupiter and Saturn are largely gaseous bodies but which possess solid cores (Jupiter does actually radiate some energy, but it is trivial compared to a star and far less than that produced by any observed brown dwarf).

To find brown dwarfs, Mayor and Queloz were using a technique which had been developed by another astronomer, Gordon Walker, working at the University of British Columbia in Vancouver. By the time Mayor and Queloz made their discovery, Walker had been searching for planets orbiting distant stars for almost twelve years. Concentrating on a group of twenty-one stars closest to our solar system, he had developed a method of detection informally known as the 'wobble' technique.

As we have seen, it is quite impossible to see objects such as extrasolar planets or even the much larger brown dwarfs orbiting distant stars because they are simply much too far away. Stars are only visible because of the vast amount of electromagnetic radiation they emit. Solid, rocky planets like Earth only reflect light from their nearby suns, and although gaseous worlds and brown dwarfs do produce a relatively tiny quantity of energy (via processes other than fusion), they would only be observable from Earth by reflecting the light from their own sun.

We can see some of the planets in our solar system with the naked eye, but a few orbit too far from the Sun to be observed without telescopes. Indeed, Uranus (which can just about be seen as a faint pin-prick of light on a clear, moonless night) was only discovered in 1781 by William Herschel. Uranus orbits our Sun at an average distance of 1.79 billion miles (2.9 billion kilometres), which is a long way off in everyday terms but only about 1/14,000 of the distance to the nearest star. The still more remote Neptune is invisible to the naked eye and was only noticed as recently as 1846 using a telescope at an observatory in Berlin. But the huge difference in distance between extrasolar planets and those at the outer reaches of our solar system is not the only consideration. Any planet orbiting a distant star would have its meagre reflected light completely swamped by the intense radiation emitted by the star it orbits. Imagine how difficult it would be to pinpoint a fire-fly hovering close to a spotlight from the distance of several miles.

So, straightforward optical observation is almost useless to astronomers searching for planets orbiting distant stars. The wobble technique works on an entirely different principle from direct optical observation. When astronomers say they have detected planets orbiting other stars, what they mean is that they have observed the gravitational effects upon the nearby stars caused by the presence of these planets, an effect which causes the star to wobble slightly from its predicted course. But how do they determine this?

Scientists measure the distance of stars using an effect called the 'red shift'. This concept originated from the work of an Austrian scientist named Christian Johann Doppler, who predicted in 1842 that if a source of sound is moving towards or away from a listener, the pitch, or frequency, of the sound is higher or lower than when the source is stationary. An everyday example of this effect is the change in pitch of the siren of an ambulance or a police car as it approaches and recedes from the listener. The Doppler effect works in an identical way with light waves, so that the colour of a luminous body changes in a similar manner – the wavelength of the light from a distant star will be longer (lower frequency) if it is moving away from us.

It was discovered earlier this century that the Universe is expanding, so every point in the Universe appears to be moving away from every other point. Because of this, thanks to the Doppler effect, light from far-off stars arrives here with a longer wavelength than it would if all the stars were stationary. This shift is called the red shift because the light arriving here from stars has been shifted further towards the red (or longer wavelength) end of the spectrum. A corresponding 'blue shift' (towards shorter wavelengths) occurs if a stellar object is found to be moving towards us.

So how is this related to the presence of planets orbiting these stars? The answer lies in the details of the frequency shift observed for some stars. Mayor and Queloz noticed that the shifts in the frequency of light from several stars fluctuated uniformly. In other words, there was a tiny fluctuation in the red and blue shifts of certain observed stars which implied that something was causing the star to 'wobble'.

A suitable analogy would be a hammer-thrower at the Olympic Games. The hammer-thrower pulls on the wire connected to the hammer and controls its path, but, although the hammer is much lighter than the thrower, it will pull on him (albeit to a very much smaller degree). It will consequently 'wobble' the thrower and, if anyone were sufficiently interested, employing accurate equipment, this effect could be measured.

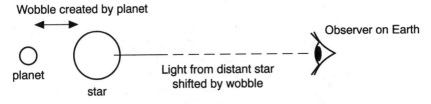

Fig. 11: How red and blue shifts show the presence of extrasolar planets.

The tug of a planet around a star is far less significant than the effect of the hammer on the thrower in the example just described. Even brown dwarfs orbiting stars are far less massive than the average star, so the effect (especially when considering a planet) is minuscule, but it is just about observable using very sensitive equipment. And, despite being a tiny effect, coupled with other techniques, it can yield a great deal of information about the star and its planets. Most significantly, Gordon Walker, who devised the technique, found that the variation in frequency of the star (the degree of wobble) is directly related to the orbital period of the planet. Using this, astronomers were soon able to paint a picture of what the first planet found beyond our solar system could be like, and where it was in relation to the star.

The first star that Mayor and Queloz found to yield evidence of a planetary system is named 51-Peg, in the constellation of Pegasus. It is very similar to our own Sun (which is a G2 star) and is designated a G3-type star, which means it is stable, around the same age as the Sun, and has a similar surface temperature.* However, that is where the similarities to our solar system end.

Until 1995, the only solar system known to the human race was ours. In our solar system the Earth is positioned third from the Sun and is one of four small rocky worlds in relatively close orbit to the Sun, which are accompanied by a set of gas giants including Jupiter and Saturn, lying further out; the whole arrangement is completed by an asteroid belt (between Mars and Jupiter); and a large number of moons orbit both the rocky and the gaseous worlds. Until 1995 this was our only paradigm, and so we could only speculate that it was a quite unremarkable arrangement. But as one of the most prolific planet-finders, Geoff Marcy, has said

---

* Stars are categorised according to the Harvard classification system. The modern version of this assigns the letters O, B, A, F, G, K and M, based upon the surface temperature of the star. O stars are the hottest, with surface temperatures in the region of 30,000K; A stars have surface temperatures of about 10,000K; G stars, such as our Sun, have surface temperatures of about 6,000K; and M stars are the coolest, with surface temperatures of about 3,000K. The numbers 1, 2, 3 etc. are a further refinement of the system, so it can be seen that 51-Peg (a G3 star) and our Sun (a G2 star) are quite similar in their chemical make-up.

recently in reference to alien races that may live on extrasolar planets somewhere out there: 'Maybe they think we're weird!'[1]

The first shock to those who held the view that our solar system was the standard model came when calculations were made concerning the mass and position of this new world. It was found that the planet orbiting 51-Peg is about half the mass of Jupiter but is positioned only 0.05 astronomical units ('a.u.') from its sun. An astronomical unit is the distance from the Sun to the Earth – 93 million miles, so the newly-discovered planet orbits 51-Peg at a distance of only 5 million miles. Furthermore, it takes only four days to complete an orbit, compared to Jupiter's sedate twelve years.

Initially, these findings were so shocking that Mayor and Queloz thought that they were observing a freak stellar system involving an exceptionally small brown dwarf somehow trapped in close orbit around 51-Peg, but this seemed unlikely as the lower limit thought possible for brown dwarfs is not less than twenty times the mass of Jupiter. It appeared that this could not be the answer, but to confirm their findings they returned to studying the fine detail of the observed Doppler shifts and to employ spectroscopic analysis to confirm or negate their hopes.

Spectroscopic analysis is another powerful tool for astronomers, as we have seen, and allows them to determine the chemical nature of a distant star. It is a staggering fact that scientists who have never had the opportunity to study physical material from any observable star are able to describe their exact chemical natures. They can do this thanks to the technique of spectroscopic analysis first devised during the 1930s as a result of Albert Einstein's work conducted on the nature of the atom.

In a paper published in 1905 (for which he later won the Nobel Prize), Einstein showed that different materials absorb or emit electromagnetic radiation in various ways depending upon their electronic structure (this is the fundamental principle behind such everyday devices as photoelectric cells, the cathode-ray tube found in televisions, and lasers). So, if the nature of radiation emitted by an object is indicative of its atomic or chemical nature, then by simply studying this spectrum alone

scientists can learn a great deal about the chemical characteris-
tics of the object. For the astronomer interested in the
composition of a star, the same principle applies. Light from the
core of the star passes through its atmosphere. This radiation
excites the electrons in the atoms of different substances in the
atmosphere, which then emit characteristic radiation. These
travel through space and are detected by the astronomer's spec-
troscope on Earth. Once processed, this information reveals the
chemical characteristics of the molecules which had emitted the
radiation in the star. This is no different to the way an experi-
menter in a lab can study the contents of a beaker of solution by
carrying out a spectroscopic analysis of the sample; the only dif-
ference is that the light from a star has to travel a lot further
than the light from a tiny lab sample.

Using this method Mayor and Queloz were able to confirm
their suspicions about the stellar system, and quickly reached the
irrefutable conclusion that the object orbiting 51-Peg is indeed a
planet and not a brown dwarf. A few weeks later, in August 1995,
they submitted to *Nature* their paper detailing the find, and then
announced the discovery publicly at an astronomy conference in
Florence the following October.

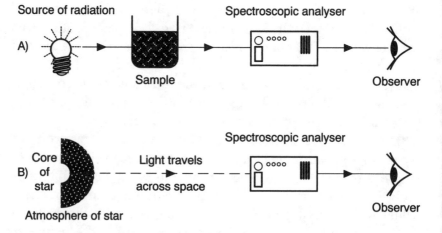

Fig. 12: Spectroscopic analysis (A) in the laboratory and (B) how
the astronomer utilises the same technique.

Not surprisingly, the response was immediate and overwhelming. Not only were the scientific community enthralled and captivated by the news, the public reaction was every bit as great as one would expect. Newspapers around the world headlined the story, a flood of features appeared in serious magazines, and sensationalised versions sprung up in a wide range of tabloids and pseudo-science periodicals. The discoverers received hundreds of phone calls, faxes and e-mails within hours of the announcement, including a wonderful e-mail message from a six-year-old American boy who wanted to know if Mayor and Queloz had visited their planet yet.

And the excitement and public interest were not the only reactions. Almost immediately, workers in the same field returned to their laboratories and began searching for other planets, an effort which produced a spate of further discoveries during the second half of 1995 and the early months of 1996.

Leading rivals of Mayor and Queloz are two American astronomers, Geoff Marcy from San Francisco State University and Paul Butler at the University of California, Berkeley, who had been working together analysing a collection of sixty stars for some seven years before the Swiss team's breakthrough. They had failed because, like almost all astronomers, they had pursued a conventional search, thinking that any alien solar system would follow the pattern of our own – small planets orbiting close to the sun and large gaseous worlds at far greater distances. They had not considered the possibility that large bodies such as the object Mayor and Queloz had found could exist so close to a sun. The findings of the Geneva team prompted Marcy and Butler to return to their data and to reappraise it in the light of this new insight.

It did not take them long to find a small collection of objects that fitted the bill, and they have since become the most successful team to locate alien worlds, with their names attributed to seven of the dozen or more worlds confirmed or close to confirmation at the time of writing. Their first find was a planet orbiting a star named 70-Virginis, an object which has since been designated as an 'eccentric' planet. About nine times the mass of Jupiter, and taking just under 117 days to complete an orbit, its eccentricity comes from the fact that it follows an exaggerated elliptical path around 70-Virginis.

The planets of our solar system all have elliptical orbits, but these paths are only slightly deviated from circular orbits. Marcy and Butler's new find travels through an ellipse that takes it to within 0.27 a.u., or about 25 million miles of its star, before travelling out to a distance of 0.59 a.u. (about 50 million miles). Within days of this finding, Marcy and Butler announced a third planet, this time orbiting 47-Ursae Majoris. This is thought to be another Jupiter-like world but with twice Jupiter's mass. It has a conventional orbit, lies about twice as far from its sun as the Earth does from ours, and takes about three years to complete an orbit.

Since these discoveries, the list of alien worlds has grown steadily. Some of these newly-located planets are considered so eccentric as to cast doubt on whether they really are planets in the conventional sense. But what they could be if they are not planets remains a mystery. As Marcy and Butler have said: 'Maybe they represent a new class that theorists haven't envisioned yet.'[2]

Recently, a new term – 'superplanet' – has been coined to describe a further set of newly-observed objects. These are all large bodies that do not appear to be brown dwarfs because their chemical composition ratified by spectroscopic analysis shows they did not form in the way stars are conventionally created (by the fusing of large quantities of gas and dust). However, they are like no other planet astronomers have ever seen, and may have formed as a result of collisions between two or more gaseous giants. For the astronomer, the most intriguing aspect of these discoveries is the growing evidence that almost all these planets lie very close to their suns (closer than Mercury is to our Sun).

Until the recent spate of discoveries, planetologists believed that there were only two types of planet: Jupiter-like worlds, the gas giants, many times larger than the Earth, orbiting the Sun at relatively great distances (in our solar system this category includes Jupiter, Saturn, Uranus and Neptune); and Earth-like worlds – rocky, solid planets, much smaller than the gas giants and found to orbit much closer to the Sun. This latter group includes Mercury, Venus, Earth and Mars.* It was never imagined possible that a large, Jupiter-like world could exist so close to a star as the objects found orbiting 51-Peg, 70-Virginis and others.

Of course it is easy to see why these odd worlds should be discovered first. The technique of detecting the wobble of a star to determine whether it is host to a planet obviously favours those stars with large objects close by, because these systems will demonstrate the most discernible gravitational irregularities. But there is still some considerable surprise in the very idea that such large worlds could exist in such close proximity to their suns – how could they have been formed, and what prevents them being consumed by the even more massive star?

To understand this we have to look at the favoured theories used to explain the formation of solar systems such as our own. According to the most modern ideas, a young sun is surrounded by a disc of dust, gas and ice. If a large 'blob' of matter coagulates in this disc and reaches a mass several times that of Earth, its gravitation field becomes intense enough to consume large quantities of gas nearby. This then forms what astronomers have called a 'protoplanet', which may then become a gas giant such as Jupiter. Meanwhile, any smaller planets that may form do so by coagulation of dust and ice to form a solid core, which then cools to produce planets such as the Earth or Mars.

According to this theory, the large gaseous bodies forming in the disc surrounding the young star are pushed outward by heat and solar winds, which are thought to be far more powerful during the early days of an embryonic star.

Since the discovery of at least four Jupiter-like planets each orbiting their stars at close range, this theory has had to be adapted. The latest scenario is that under appropriate conditions some gas giants can begin to spin rapidly and spiral in towards

---

* The characteristics of Pluto remain something of a mystery, but it is thought to be an icy world with a mean density close to that of water. It has a fairly eccentric orbit (although nothing compared to some of the new alien worlds) which brings it within the orbit of Neptune at certain times.

their stars, obliterating any terrestrial-like planets in their fiery wake. Some of these may plunge into their stars, but others obviously do not. This is explained by the fact that as the giant planet approaches its star, complex gravitational interactions transfer some of the momentum from the planet's rapid spin into energy in the form of orbital motion, and the planet takes up a stable orbit close to the star.

For those looking for answers to the question of whether there is life on other worlds, the recent findings of astronomers are a double-edged sword. On the one hand they show that there are plenty of solar systems other than our own. However, the downside of this is the nature of the solar systems that have been found so far.

Of the dozen planetary systems confirmed to date, the majority are eccentric in one form or another; that is, they do not conform to accepted models for the construction of solar systems. Fundamental to the old standard model is the idea that only small, solid planets can be found close to suns; this has been shown to be quite false. But beyond this is a growing suspicion (backed up by impressive research) that certain characteristics of our solar system allow it to remain stable over long time periods. One important example is the recently devised theory that gas giants act like 'planetary shepherds', protecting the inner worlds upon which intelligent life could form by 'mopping up' dangerous objects such as life-threatening meteors and comets.

By this reasoning, if life did manage to form on planets in systems such as that found around 51-Peg (despite the fact that the gas giant[s] of this system span out of control and then took up residence near the star), then, because there do not seem to be other giant worlds at Jupiter-like distances in this system, the inner worlds would not be protected as theory suggests they have been in our solar system.

Of all the systems detected so far, the one that seems initially to be similar to our own is that which may exist around 47-Ursae Majoris. This is a star that fits approximately the same category as our own (it is a G0 star) and the detected planet is within the size range we would expect for a gas giant – between two and three

Star	Spectral type	Planet mass (Jupiter = 1)	Average distance from star (a.u.)	Orbital period
51-Pegasi	G3	0.46*	0.05	4.2 days
Rho¹ Cancri	G8	0.84*	0.11	14.7 days
Tau Bootis	F6	3.87*	0.05	3.3 days
Upsilon Andromedae	F7	0.68*	0.06	4.6 days
HD114762	F9	9.0*	0.34	84 days
70-Virginis	G5	6.5*	0.43	116.6 days
16 Cygni B	G2	1.5*	1.72	2.2 years
47-Ursae Majoris	G0	2.3*	2.1	3 years
Lalande 21185	M2	~0.9	~2.2	~5.8 years
Lalande 21185	M2	~1.1	~11	~30 years
Rho¹ Cancri	G8	~5	~5	~20 years
Upsilon Andromedae	F7	?	?	~2 years
Lalande 21185	M2	?	>11	>30 years

* = minimum mass

Fig. 13: Members of the planetary zoo.

times the mass of Jupiter. Most importantly, it is in roughly the right place, about 200 million miles from 47-Ursae Majoris, compared to Jupiter's orbiting distance of about 480 million miles, and its orbit is also non-eccentric.

Another possibility is the system of the star Lalande 21185. This has been detected by George Gatewood of the University of Pittsburgh. Lalande 21185 is a red dwarf star that happens to be one of our closest stellar neighbours, lying only 8.25 light-years from Earth. Gatewood has detected what he thinks are two gas giants orbiting Lalande 21185. Both are almost exactly the size of Jupiter, one positioned a little over 2 a.u. from its sun (equivalent to the orbit of the asteroid belt in our system), the other at 11 a.u., equivalent to a distance a little beyond the position of Saturn in our home system. However, Gatewood uses a different detection technique from most other astronomers, a method which does not immediately give accurate values for the mass or orbital distance, and his findings are yet to be confirmed.

So what do these discoveries represent in our search for life on other worlds?

It appears that life is far less likely to evolve in solar systems containing Jupiter-like planets close to their stars than it is in systems similar to ours. The reason for this is the instability generated by planetary dynamics early in the evolution of such a system. Yet there is a caveat to this, which is that if the changes occurred early enough, there would be time for the planetary system to settle into a stable state before life took a foothold. After all, there are several theories which suggest that our solar system has been the site for cataclysmic upheavals early in its existence. It could be that at least one of the systems we have so far detected settled into a regular pattern of planets orbiting uniformly, and at the 'correct' distances, several billion years ago. This, as we have seen, allows plenty of time for life to evolve, just as it has on our own world.

However, the downside to this is the strong possibility that such systems require at least one other gas giant occupying an orbit far from the sun, acting as a shepherd for the inner worlds where life could have a chance of establishing itself.

There are two other factors to consider. One is the small sample of stars analysed so far. Using the wobble technique and others, astronomers have gathered data on no more than a few hundred of the estimated 400 billion stars in our galaxy alone. As I discussed in some detail in Chapter 4, there are many steps to consider between the establishment of a planetary system and the evolution of contactable intelligent life. Yet we may with some confidence survey the figures and say that it is likely that the first prerequisite – the presence of alien solar systems – has been met. Of the planetary systems found and confirmed, one looks as though it could be a system not completely different from ours, and Professor Gatewood's researches may soon double that figure. One out of twelve systems is an encouraging result.

The second factor to contemplate is that the techniques currently at the cutting edge of research into planetology only allow for the detection of large, Jupiter-like gaseous worlds. This technology will gradually improve in the coming years, and we may soon be in the position to detect smaller rocky worlds orbiting distant stars. When this is achieved, astronomers will have to construct models describing the planetary dynamics of the systems they find. It is almost certain there will be many surprises in store, transforming our image of how solar systems should be.

As with many of the recent developments in astronomy and biology discussed in earlier chapters, what has been discovered in the realm of extrasolar systems may be interpreted as optimistic or pessimistic. The enthusiasts of extraterrestrial life point to the fact that research has shown there to be a good many worlds orbiting alien stars, and that even with the relatively little work that has been conducted so far we have obtained encouraging results – the possibility of at least one sol-type system. The sceptics highlight the tremendous obstacles confronting the formation of solar systems that have a chance of possessing the correct dynamics to support life, and the seeming rarity of systems comparable to ours, which, according to the bleakest predictions, appears to be the only type capable of allowing intelligent life to evolve.

However, unlike many of the other questions facing both the enthusiasts and the sceptics, the hunt for extrasolar worlds is one area of the search in which we can be proactive. And the chase is only just beginning, with the human race at last taking the first tentative steps to search beyond our own solar system using powerful telescopes and probes. As one of the leading workers in the field, Paul Butler, has said: 'Within the next couple of years, I expect the discovery of systems with multiple planets and the discovery of a planet that is as small as ten Earth masses. We've literally just started scratching the surface, and we're going to find many things that we didn't expect to find.'[3]

7

THE CHASE

'The fair breeze blew, the white foam flew,
The furrow followed free;
We were the first that ever burst
Into that silent sea'

SAMUEL TAYLOR COLERIDGE,
'THE RIME OF THE ANCIENT MARINER'

The desire to broaden horizons is a fundamental human trait. The drive to learn, to explore, to discover, is an integral part of our mental framework, and for this reason, there is no doubt that if we survive as a species, then our future lies in space.

Our early human ancestors probably left Africa to populate the world thanks to the drive to find fresh resources, particularly food. The first explorers who ventured forth from Europe to the furthest reaches of the planet, to the Americas and Australia, made this effort for similar reasons – the search for gold and spices: resources of a different kind. They were also motivated by political pressure – the desire to beat their rivals to the gold and the spices. It was for this reason too that humans finally succeeded in venturing beyond our world to take the first tentative steps into the Big Out There.

The people who made this possible – the British, American, European and Russian scientists, the brains behind Mercury, Gemini, Soyuz, Apollo and the others – were motivated by the desire to know, to explore, but these were not the men and women who made it possible on another practical level; it took others of different persuasions to produce the finance for such an ambitious undertaking. This came from politicians, the governments of the United States and the former Soviet Union, men and women who were determined their nations would be first in space, first on the Moon, first on the front page of the newspapers of the world. Our passage into space was facilitated by the vagaries and insecurities produced by the Cold War. Indeed, the Cold War was an essential element of the Space Race.

To get humans into orbit or to the Moon is one thing, but to colonise the planets of our solar system and to reach further afield, to the stars beyond, to live permanently in an alien environment, to raise new generations of people, humans who may never see Earth, is an entirely different proposition. This effort will certainly be precipitated by political ends, but eventually it must become self-sustaining.

NASA is really the only important space agency in the world today. The Russians are chronically short of funding and have been reduced to leaving their cosmonauts in space for longer than planned because they cannot afford to bring them home. Britain forsook its long-term role as a member of the space club in the 1960s, but remains a key player in the development of technologies and research. The European Space Agency is growing but remains a very junior sibling of NASA, and the Japanese are gaining ground but have almost half a century of development to catch up.

At present, NASA survives by the skin of its teeth, relying upon the impetus of new breakthroughs and vulnerable to the whim of fluctuating public opinion. It has a vast annual budget (£8.2 billion for the current year) but this can be wiped out at a stroke if the US government chooses, or boosted if something sensational happens, so the agency continues to walk on egg-shells. The recent indication that there may once have been life on Mars (and that there is a slim chance there may be life there still) has boosted interest in space travel and this has been reflected in NASA's budget, but money remains the primary barrier to the development of a real space culture. Money stops us going to Mars, stops us building bases on the Moon and prevents us from developing environments for long-term space exploration. This is particularly galling to the space enthusiast, weaned on the great hopes of the early Space Age, when it looked as though humans would be on the surface of Mars by the 1990s and Moon bases would be holiday resorts by the end of the century. What makes this even more frustrating is the fact that the NASA budget is approximately equal to what the US government spends on defence every *two weeks*.

The factors that have allowed humans to achieve what they have managed so far in space will not sustain any serious efforts to reach the planets of our solar system, let alone the distant stars. So far we have been motivated by political drives, the desire to learn, the hope of finding life on other worlds. To make space travel work, we will have to add to these ambitions the attractions of Mammon. When space becomes a net profit-making venture rather than a massive loss-making one, then we really will have entered the Space Age – one that will be self-perpetuating and long-lived.

The huge boost to the NASA budget created by the Apollo Moon landings gave it a mandate to develop the shuttle programme and a series of unmanned missions to the planets of our solar system. For the most part, the discoveries these brought forth sustained the space agency, but some almost killed off the interest of the money men. One such programme was the Viking missions to Mars.

The primary purpose of the two Viking probes that landed on Mars during the autumn of 1976 was to find extraterrestrial life. They were the first attempts by humans to find life out there. Each lander contained two simple sets of experiments. The first set consisted of three biological tests which involved exposing a scoop of soil to different chemical reagents and testing the gases produced. All three tests gave positive results, that is, they all showed the production of gases indicative of biological processes.

When these results were processed by the mission controllers back on Earth they produced an instant buzz of excitement, and within hours the news that there might be life on Mars was leaked to the press, obliging the team to publicly deny the news. One of the team leaders, Norman Horowitz, had to go on record with the comment: 'I want to emphasise that we have not discovered life on Mars – *not*.'[1]

With a growing sense of anticipation, the second batch of experiments were started. These involved analysing the Martian soil samples with a device called a gas chromatograph mass spectrometer (GCMS), which studied the chemical composition of the sample. To everyone's amazement, this set of tests showed the very opposite of the first batch – according to the experiment, there was absolutely no trace of any form of organic molecule in the sample.

It would have surprised researchers if the second test had shown no sign of biochemical materials such as complex amino-acids, but to show that there were no organic materials *at all*, not even simple organic structures, was almost unbelievable. Confused, the team then turned their attention to the second lander, Viking 2, which had arrived on the Martian surface a few weeks after its predecessor and contained exactly the same package of tests. The first set of three experiments were conducted and gave exactly the same results as those carried out by the rudimentary laboratory aboard Viking 1 – they showed the undeniable presence of gases indicative of biological activity in the sample. Viking 2 was then instructed to turn to the second experiment, and the material was passed through a GCMS. The team sat back and waited anxiously for the results. Some hours later, the data came through. To their further surprise, this second test on the second lander showed exactly the same result as the second experiment on Viking 1 – no sign of any organic materials whatsoever.

For the NASA team, this conflicting set of results was more frustrating than exciting. If such results were obtained in a lab on Earth, they would immediately prompt further investigation to determine the reason for the contradictions. But here, tens of million miles from the action, nothing could be done. The landers were simple devices that were capable only of landing and carrying out a set of pre-programmed, elementary tests; there was no scope to investigate further.

Today, more than two decades on, the debate about the results of the Viking tests remains unsettled. There are advocates for the argument that the results showed clear signs of life in the Martian soil; perhaps primitive bacteria, and there are those who proclaim a diametrically opposite view. Sceptics believe the first set of tests did not indicate life at all, but represented complex inorganic reactions, possibly involving chemical groups such as peroxides and superoxides which might replicate the behaviour of biochemicals.

The outcome was not only frustrating for the scientists working on the project, it injected a feeling of deep pessimism into the entire space programme, and attention immediately shifted from Viking and the search for life to more practical work such as that being conducted on the shuttle programme and the deep space probes (such as Voyager) sent to the outer planets. Mars was left alone, shrouded in an atmosphere of disappointment.

Fortunately that sense of pessimism and frustration did not last for ever, and we have now returned to the red planet after a hiatus of over twenty years. Three missions to Mars were launched towards the end of 1996, each taking advantage of a rare window of opportunity. It has been calculated that using present technology and working within the restraints of tight budgets, craft can only be sent to Mars during a period lasting six to eight weeks every two years. The reason for this is simply down to planetary mechanics. Approximately every two years, Mars and the Earth are aligned so they are both on the same side of the Sun, which means that the distance between the two planets is as small as it ever becomes – something in the region of 35 million miles. At other times the distance between the two worlds grows larger, to a maximum of 60 million miles, which would mean that the mission would take considerably longer.*

Three missions were launched, but only two survived to reach Mars. The first to leave was NASA's Mars Global Surveyor, which took off at the beginning of November 1996 and arrived in

---

* Mars has a perihelion (closest point to the Sun) equal to 128 million miles (207 million km), and an aphelion (furthest point from the Sun) of 154 million miles (249 million km). The Earth's orbit is almost circular, the distance from the Earth to the Sun fluctuating by only 1.5 million miles from a mean of 93 million miles.

Mars' orbit in September 1997. Little over a week later a Russian probe, Mars 96 (the largest and most ambitious of the three projects), failed to leave Earth orbit en route to Mars after its engines malfunctioned. Finally, two weeks later, on 4 December, a third mission, NASA's Mars Pathfinder, was launched successfully and landed at a site named Ares Vallis on 4 July 1997, several weeks before the Mars Global Surveyor arrived in orbit.

The loss of Mars 96 was a tremendous blow, not only to the Russians but to the teams of scientists around the world who had contributed experiments to the programme and who had spent years developing systems and investigative packages for the mission. There are serious suspicions that budgetary restraints were responsible for the loss. These follow claims that final tests on the craft were halted because of cash problems, and had to be conducted on an inadequate simulator.

So what is the objective of these missions? As the name implies, the Mars Global Surveyor is a probe that studies the surface of Mars from orbit. It is equipped with still and video cameras, instruments to study the atmosphere of Mars (including sophisticated spectrometers) as well as devices to analyse the magnetic field of the planet. It follows a near-polar orbit and dips as close as 378km above the Martian surface.

This probe offers astronomers invaluable insights into the geology of the planet and is beginning to produce detailed maps of the surface, maps that will be essential for future missions. However, for those interested in the search for extraterrestrial life, the outcome of the Mars Pathfinder mission was always the more immediately significant of the two, and captured the public imagination in the summer of 1997.

Pathfinder hit the Martian atmosphere at a speed of 16,000mph and was slowed by a set of retro-rockets and parachutes, so that by the time it reached an altitude of 65 feet it was travelling at only a few miles per hour. When it reached this altitude, giant airbags inflated to cushion the final stage of the descent. The Pathfinder probe then bounced along the Martian terrain until it righted itself, base-down. With the landing completed successfully, all systems were checked and the probe was instructed to deploy a tiny solar-powered rover called 'Sojourner', which operated totally independently from the lander and was remotely-controlled from Earth.

Sojourner weighs little more than 11kg, but it is packed with equipment and has been specially designed to cope with the Martian terrain. It has six wheels and a sophisticated sensor device which detects the presence of obstacles and determines the best way around them. At the time of writing, some six months after the landing, Sojourner is still surveying the surface. It carries cameras, X-ray spectroscopy equipment to analyse soil samples, and a set of experiments to determine whether there is simple life present in the soil. These experiments are far more sophisticated than those aboard the twenty-year-old Vikings, and have been designed to reduce to a minimum any ambiguity about whether a sample contains life or is just exhibiting odd chemical reactions.

So far we have had some spectacular pictures from the Martian surface and a continuous stream of information from the lander which is currently being processed by the scientists at NASA.

Some enthusiasts in the debate about extraterrestrial life argue that the Viking probes were just unlucky, that they picked particularly barren samples. Unfortunately, this does not stand up to close scrutiny. Bacteria exist almost everywhere on Earth. If a scientist were to analyse soil from any part of the globe, or even to test samples dredged up from the seabed or scraped from the inside of pipes at a nuclear reactor, they would be likely to find some form of simple organism. Life on Mars may eventually be found deep underground, but suggestions that the probes 'missed' the presence of life are quite untenable.

Pathfinder is undoubtedly a superb piece of space engineering and will hopefully answer some of the questions opened up by the discovery of ALH84001 and the mystery surrounding the Viking findings, but even this is no substitute for studying the real thing here on Earth. NASA plan to send similar pairs of probes every twenty-four months using the regular flight window, but the next step in the search for life on Mars will come early in the twenty-first century when, budgets permitting, a probe will be sent to Mars to collect a sample and return it to Earth. The scheduled date for this project is currently 2005, and hopes are running high that the enthusiasm generated by recent discoveries will maintain the momentum to make this dream a reality. If this is achieved, it may solve the question of whether life still exists on Mars once and for all.

Beyond these voyages to Mars, NASA has initiated an ambitious programme of investigations of the other planets in our solar system. There are currently two important missions to the outer worlds. Galileo, launched in 1989, entered the atmosphere of Jupiter in December 1996 and has been investigating the Jovian moons at close quarters, offering new insights into the potential for life to exist on these worlds.

Recently a new ethos has prevailed at NASA – the principle of 'faster, cheaper, better'. This way of thinking is best seen in a new generation of microprobes (some of which weigh only a few hundred pounds) destined for the farthest corners of the solar system within the next two decades. The latest is a probe called Huygens (named after the influential seventeenth-century Dutch scientist), which is on course to land on Saturn's largest moon, Titan, in 2004. Beyond this, there are plans to return to Venus, to send a probe called Hermes to Mercury and for a $250 million scheme to undertake a twelve-year-journey to Pluto to explore this, our furthest neighbour, and its moon Charon.

These undoubtedly exciting projects will increase considerably our knowledge of our immediate environment in space. But the big question is: when will humans venture to other worlds?

There are those who argue there is no need to send humans to other planets, that all exploration and even mining and exploitation of resources can be conducted by machines controlled from the Earth. This is certainly cheaper and easier, but to most, it is not in the true spirit of frontier exploration. We are creatures who need to be at the centre of the action. For us, probes may be invaluable today – in financial terms they are really the only option – but eventually there can be no substitute for actually being there. So what serious proposals to get humans to other planets are under consideration?

Almost since the end of the Apollo missions, NASA and other

space agencies have been working through a master-plan for planetary exploration. Instead of pulling off what at root are technological tricks stimulated by politics, such as the trips to the Moon, they are concentrating on a step-by-step process that will get human beings to the planets and beyond.

The Russians lead the world in research to investigate the effects of staying in space for long periods, and hold the record for the longest unbroken space mission. For at least fifteen years they have been concentrating on solving the problems that long trips beyond Earth cause the human metabolism and the human psyche. The original motivation for this was again political. After the Americans stole their thunder by getting to the Moon first, the Soviets concentrated on Mars. The first step along this road, therefore, is to develop strategies to deal with a round-trip estimated to take some three years (including a lengthy stay on Mars).

Another area in which the Russians have led until recently has been the construction of space stations in near-Earth orbit. The Mir space station was the first of its kind and is still in operation (just) and has been manned constantly since it was launched fourteen years ago. Space stations are an essential stepping-stone to the stars. It is impractical to start a mission with a single launch from Earth. The craft that gets us to Mars will have to be exponentially more powerful than those used in the days of Apollo. It will have to carry enough fuel for a round-trip of some 125 million kilometres (as opposed to about 800,000 kilometres to the Moon and back); the crew would probably number a dozen instead of three; and the stay on Mars will run into weeks or perhaps months. The obvious solution then is to construct a large interplanetary craft in Earth orbit, for which a fully-manned and fully-equipped space station is a prerequisite.

These two elements – research into long stays in space and the building of a space station – have been pioneered by the Russians, but the third essential ingredient is a sophisticated Earth-to-space transport system – a cargo carrier. NASA's shuttle and its descendants will act as the hauliers of materials from which a large space station will be constructed; and from here the first spaceship to Mars will be built and launched.

Plans for this are already well under way. The first components
of the international space station, co-financed and built by the US,
Russia, Europe, Japan and Canada, are due for launch this year,
and it is hoped that the station will be completed and permanently
manned early in the next century. The project has been hit by
mounting costs and a succession of budget difficulties (the
Russians have delayed things due to cash problems), but it is seen
by all involved as the crucial next step in human expansion
beyond our world.

A manned mission to Mars is still little more than a dream, but
detailed plans are on the drawing board, and the space station, the
shuttle and the Russians' experiments are all essential elements
and have reached an advanced level of development. This ground-
work is expensive and fraught with technical problems, but at least
things are moving forward, even if they are not progressing at the
speed hoped for in the early days of space exploration. Currently,
the earliest date for a manned mission to Mars is 2015–20.

As the earliest preparations for such a trip have been progress-
ing and the infrastructure has been designed and put into place,
there have been other changes on Earth. At the time NASA and
the Russian space agency were planning their strategies for the
conquest of Mars, the world's two superpowers were engaged in a
largely bloodless but nevertheless destructive Cold War. None of
us can predict how the world will be in the first decades of the
twenty-first century any more than we could have foretold the
end of communism, but at least there are now hopes that the first
people to set foot on Mars will truly be there as representatives of
all humanity.

And what can the first explorers of Mars hope to find there?

The planet is entirely inhospitable. It is not a fiery waste such
as Venus or Mercury, nor is it a totally uninhabitable world com-
posed largely of methane and hydrogen, such as the gas giants, but
humans could not live there without relying upon advanced tech-
nology because of the intense cold and the almost complete
absence of oxygen. The first explorers would have to wear space
suits all the time while on the surface of the planet. The temper-
ature at the equator in high summer rarely nudges past freezing

point and the atmosphere is made up almost entirely of carbon dioxide with traces of nitrogen and noble gases. Furthermore, atmospheric pressure is less than one hundredth of Earth's so there is no chance of finding surface water on the planet.

Basically, Mars is an airless, freezing desert that makes the Sahara or deepest Siberia appear an oasis by comparison. Nevertheless, this will be our first port of call and a staging-post for exploration of the rest of the solar system. However, just getting there and building a simple temporary base requires overcoming a plethora of technological and practical problems.

The technology of interstellar travel is dealt with in the next chapter, and as you will see, the difficulties inherent in travelling between the stars are formidable. Although inter*planetary* travel is a far less daunting proposition, it is by no means easy. The technological challenges include developing ways to recycle all forms of waste products; overcoming the natural muscle-wastage that occurs during long flights (as experienced by cosmonauts on Mir); defending against deadly cosmic rays which flood space beyond our atmosphere; developing transport systems and power sources on alien planets; finding ways to build safe habitats, techniques to mine the planets and asteroids and the production of systems by which these resources may be returned safely to Earth.

Although these are difficult problems to solve, within a few centuries they will be overcome and humans will travel around our solar system and eventually build homes in every corner that could possibly sustain us. But the other side of the equation is the enormous cost associated with such an undertaking.

The exploration and material exploitation of the solar system is of course far beyond the resources of individual nations, and may well be beyond the ken of a group of governments. Even if the world's powers continue to co-exist peacefully and co-ordinate their efforts, as they are already starting to do, the trillions of pounds involved in developing technologies to travel quickly and freely throughout our solar system, to extract minerals, to live in space for long periods and to communicate rapidly, will have to be raised via commerce.

Convincing businessmen to invest in research and development is never easy, but recently a few far-sighted entrepreneurs have started to set up the first independent space agencies which, it is hoped, will eventually rival the government-funded institutions. Two of these early pioneers are Geoffrey Landis, a senior research associate at Ohio Aerospace Institute, and his partner, David Burkhead, who have designed a craft called SpaceCub, the first DIY spacecraft. With a proposed price tag of around $500,000 it comes far cheaper than a single shuttle ride, and it is reusable. Powered by Soviet engines left over from the Cold War and now available 'off the shelf', it is designed to carry a single passenger on a 1,500-kilometre round-trip, reaching a velocity of around 7,000kph for about 20 minutes.

More ambitious are the plans of a US Air Force Major, Mitchell Burnside, who has designed a single-stage-to-orbit craft along the lines of NASA's own shuttle, called Black Horse. This is a genuine orbital-altitude craft that travels at Mach 15, about ten times the speed of Concorde. Developing the prototype will cost $150 million, which sounds prohibitive, but as Mitchell points out: 'It is no more than the cost of a single launch of NASA's workhorse launcher, the Titan IV, paid for regularly by the space agency.'

Taking things several stages further are the Artemis Foundation, a non-profit-making lobbying group based in Houston who plan to establish a Moonbase by the year 2003. They intend setting up a commercial organisation charging for civilian flights to the Moon, where passengers will stay at Moonbase Artemis. Their target date of 2003 might appear a little optimistic, but they are quite serious. They plan eventually to go public with their Lunar Resources Corporation and to raise funds from industry supported by profits from merchandising – anything from videos to models of the Moonbase – and by selling the film rights to the highest bidder in Hollywood.

The Artemis Foundation may seem naïve, but the prospects for independent space agencies look good. The main thrust of this is the idea that, in order to explore, there has to be another angle that makes money. One of the key ways envisaged by these entrepreneurial pioneers is to raise capital from tourism.

Tourism is one of the biggest industries on Earth, and space is the inevitable new frontier for those with adventurous spirits and deep pockets. Already the Japanese are planning a 'hotel' module for their section of the international space station, and they hope to be entertaining the first fee-paying guests there by the end of the first decade of the next millennium. Space tourism was a dream long before the Moon landings, but now the shuttle has settled into an almost routine schedule of trips into Earth orbit and construction of the space station is about to begin, it is being taken very seriously.

Once a commercial system of exploration, resource exploitation and tourism is established, the only barrier to human expansion into the solar system will be technical – ways to survive in the most hostile environments imaginable. So what will this mean to those searching for extraterrestrial life?

Professional opinion about the possibility of finding life in our solar system is slowly shifting. There was a time when the very idea of life on other planets within our own star system was thought ridiculous, but recent discoveries have generated optimism. Although many scientists are justifiably sceptical about the true value of these discoveries, it is not difficult to find others who believe that life may still exist on Mars. Viewed optimistically, it may not be long before our suspicions about Mars and moons such as Europa are confirmed, but it will be quite some time before we can actually travel to these places ourselves. Until this is possible we will have to rely upon robots, rovers and probes to provide us the experience vicariously. But, gradually, we will venture to the planets and establish outposts there. These will develop into communities and then into new homes for the expanding human race.

Predicting time-scales for something so ambitious is almost impossible, but it would be reasonable to suppose that within fifty years humans will have established permanently-manned bases on the Moon and Mars. Within a century humans will have begun exploring every planet of the solar system and outposts will exist in such seemingly hostile environments as the Jovian and Saturnian moons, some of the larger asteroids, and possibly even Venus.

One of the primary motivations for these efforts would be the return in resources, because by then our own may be at alarmingly low levels. Our own Moon, as well as the far distant asteroids (which lie largely between the orbits of Mars and Jupiter) are known to be rich in minerals, and within the next decade we will learn more about the geology of Mars, the moons of the gas worlds and beyond. These may turn out to be vast repositories for all the resources humankind will need for the next millennium, an era in which we might develop embryonic technologies to allow us passage to the stars.

However, professional opinion concerning the way we will one day make use of the entire solar system varies. Isaac Asimov once said: 'If anyone thinks that the important reason for exploring space is to find outlets for our expanding population, let him think again . . . within five thousand years, at our present rate of increase, the total mass of flesh and blood will equal the mass of the known Universe.'[2]

To some degree this is playing with numbers, and there are all sorts of factors that would stop such an absurdity, but the management of space travel will be a major political and sociological issue in the coming millennia, as it almost certainly has been for other civilisations. One pundit, the writer John Lewis, has suggested that within 1,000 years the population of the solar system will level off at about ten million billion people (two million times the population of the Earth today), at which point we might be capable of interstellar travel and see a mass exodus to the nearest stars.[3]

And this expansionism will be fraught with new triumphs and new problems for humankind. Dilemmas will arise and be confronted which so far only science-fiction writers have imagined and investigated. Not least of these will be the phenomenon of bifurcation.

For almost a million years there has been only one human species on the planet. But as we take the next technological leap, the creation of a true Space Age (perhaps the biggest step since we emerged from the sea), new varieties of humans will evolve, humans born on distant worlds who may never visit the Earth. These 'Martian humans' could be the first aliens we ever meet face to face. But how is this possible? How could humans ever be independent of Earth? Surely we will only travel on 'missions' whether exploratory or commercial?

This is certainly true in the short term. For centuries to come, all space exploration will remain in the hands of pioneers. Mars and the moons of Jupiter will be the new Wild West, colonised by the brave descendants of those who made every nook and cranny of the Earth their home. But what happened when Europeans colonised the New World? Within a few generations the descendants of the Pilgrim Fathers participated in the War of Independence. Americans did not evolve into a new species because the environment of the American continent is almost identical to that of Europe; and although transport between the Old and New Worlds was difficult, relative to interplanetary travel it was simple and readily managed. If we transpose this scenario on to a future where Mars is the real New World and generations of Martians are born and live their entire lives on an alien planet, it is clear that the analogy breaks down, and that the human race will bifurcate into an Earth-bound species of human and a growing group of human extraterrestrials.

Gradually the descendants of those who establish permanent homes on alien worlds will evolve characteristics which make them better adapted to the natural environment of their home world. Future Martians of Earth-human ancestry may perhaps develop lungs that can work more efficiently, so they would not need to rely entirely upon oxygen tanks strapped to their backs. They may also develop systems to adapt to the cold and to resist the intense radiation experienced on the almost airless Martian surface.

If these 'Martian humans' travelled to Earth they would find the environment almost as hostile as we would if we travelled to Mars. The earth's gravity is far more powerful than that of Mars, so these visitors would feel incredibly heavy and sluggish. The air would be too rich in oxygen for them, and most places on Earth would feel unbearably hot.

Although bifurcation is the most likely outcome of interplanetary travel, there is one alternative – the possibility that we will eventually be able to make other planets similar to Earth. This process is known as 'terraforming', and although the concept of changing the environment of an entire planet sounds incredibly far-fetched, the idea is taken seriously by many scientists.

The terraforming of Mars is perceived as the standard model for such an idea. In order to achieve this, Mars would have to be provided with an atmosphere which would in turn raise the surface temperature and, according to theory, kick-start natural mechanisms that would then take over, allowing the planet to produce a stable, self-sustaining ecosystem not unlike our own. Advocates of this idea include such far-sighted thinkers as Arthur C. Clarke, who wrote an entire book on the subject called *The Snows of Olympus*; James Lovelock, the creator of the Gaia principle; and the late Carl Sagan, who considered the subject in some detail in his ground-breaking book and TV series, *Cosmos*. It also has the support of several eminent biologists and space engineers at NASA.

Naturally, the technologies involved are way beyond anything we could conceivably develop today, but they are not beyond our imagination; and, crucially, none of the concepts involved runs up against the laws of physics – terraforming may not yet be possible, but it is certainly feasible, at least in theory.

One way in which this seemingly impossible task could be accomplished is to cover a large area of the planet with carbon (simple soot). The surface of Mars is already relatively dark and most of the light reflected by the surface comes from the polar ice-caps. If these were covered with a thin layer of carbon, calculated to weigh a few million tons, this would create the dual effect of raising the temperature and releasing vast quantities of carbon dioxide (and possibly water) from the poles.

A few million tons is a considerable quantity of carbon, but there is already a plentiful source in orbit above the planet. One of the two tiny moons of Mars, Phobos, contains several billion tons of carbon, so in principle it should be relatively easy to coat the polar ice-caps.

Another suggestion is to place into orbit vast mirrors, equivalent to the entire surface area of the planet. Although this in itself would seem an impossible task, the mirrors would need to be only a few atoms thick and perhaps made of specially prepared foil. Their effect would be to double the amount of solar radiation reaching the Martian surface and to create a corresponding rise in the mean surface temperature.

Another idea is to trigger a special kind of 'slow-burning' of Phobos, which would produce a source of heat equivalent to about one-tenth of that reaching the Martian surface from the Sun. Given a few centuries, this would raise the temperature of the planet to something close to the mean ambient temperature of the Earth.

These systems might provide an equitable surface temperature, but other techniques would have to be used to provide a breathable atmosphere. At present, the concentration of oxygen in the Martian atmosphere is about one-hundredth of that needed for human explorers to move and work comfortably. To create an atmosphere, gigatons of oxygen and nitrogen would have to be found and trapped within the planet's gravitational field.*

There may be vast amounts of water trapped under the Martian surface, but even the most optimistic estimates suggest this would only provide a fraction of the oxygen needed if it was to be released and broken down into oxygen and hydrogen. Much of the free oxygen that may have once been abundant on Mars was long ago trapped in carbonates in the soil and carbon dioxide in the atmosphere. This would be very difficult to liberate, but one far-out suggestion is to attempt to release the necessary oxygen by using millions of high-yield thermonuclear bombs. This would not be easy, especially if we consider the fact that by the time this scheme could be put into operation there may already be a substantial population of humans living on Mars.

A seemingly bizarre alternative is to bring the gas to the planet. This could be done by capturing an asteroid, many of which are thought to contain vast amounts of frozen water. After being guided into a suitable orbit, the asteroid would enter the Martian atmosphere and break up, showering the planet with ice. This

---

* A gigaton is 10^9, or one thousand million tons.

would vaporise and then decompose to produce hydrogen and oxygen. Another idea is to guide into orbit a comet (also largely composed of ice) which could be manoeuvred into a gentle descent into the atmosphere, where it too would decompose.

These ideas may appear to be pure science-fiction, and at the moment they remain precisely that, but the overall principles are contained within the laws of physics; it is merely a question of scale. The technologies required to produce such a radical change in the environment of a planet lie perhaps thousands of years in the future, and more time (several millennia) would be needed to actually bring about such changes once set in motion. But one day, if we wish it to be so, Mars will be a green and blue world, like our own world today. Humans and 'Martian humans' will walk through fields and wade in rivers created by the intervention of mortals, our own descendants. And, for all we know, we too walk through fields and wade through rivers here on Earth that might once have been 'seeded' and set on a self-sustaining path by beings not so very different from us.

8

WILL WE EVER REACH THE STARS?

*'Space is big, really big. You just won't believe how
vastly, hugely, mind-bogglingly big it is. I mean you
may think it's a long way down the road to the chemist
but that's just peanuts to space'*

DOUGLAS ADAMS, *THE HITCHHIKER'S GUIDE
TO THE GALAXY*

Indeed, the Universe is a very, very large place. In fact most of us cannot begin to conceive of just how vast it is, and many people muddle up ideas about interplanetary distances, the distances between the stars and the unimaginable chasms between galaxies.

Often, commentators refer to our own solar system as 'outer space'. This is really nonsense. As the name implies, our solar system is a local family of planets all of which orbit the Sun at different distances. Mercury is the closest to the Sun, at an average distance of 45 million kilometres, and Pluto is the furthest, some 4,500 million kilometres from the Sun.* The Earth, along with

---

* To put this into some perspective: it was recently announced that communications with Pioneer were being shut down because the information it was transmitting was not worth the cost of running the receiver. The unmanned satellite Pioneer 10 is currently travelling at 28,000mph and is twice as far from the Sun as Pluto. It has taken twenty-five years to get there at this speed. Signals from Pioneer travelling at the speed of light take nine hours to reach Earth.

Mercury, Venus and Mars, is considered an inner planet, and Jupiter, Saturn, Uranus, Neptune and Pluto are the outer worlds.

Now, talk of hundreds of millions of miles sounds impressive, and such distances are rather stupefying for us today, but the scale of our solar system is as nothing compared with the figures involved in interstellar travel. Let's consider an analogy. Picture a bubble of air, say three or four centimetres across. Now imagine that this bubble is home to a group of tiny creatures who live and die within the bubble. Now picture this bubble floating down stream to the river and on to the sea and finally ending up in the middle of the Pacific Ocean. This could be a model for the size of our solar system (the bubble) in our galaxy – the Milky Way (the entire Pacific Ocean). Now imagine those incredibly small creatures (whose entire solar system exists in the bubble) trying to reach another bubble somewhere else in the ocean, one that is say, 5 kilometres away. This would be the equivalent to us travelling to the nearest star, which is about 4 light-years from Earth.

The point of this analogy is to illustrate that interstellar distances are incomparably larger than interplanetary distances, and that the gulf between galaxies and then families of galaxies (or clusters) and then the size of the Universe as a whole is proportionally colossal.

The facts of interstellar travel (a technology which hopefully we will one day master) all revolve around distance, time and power. Because the distances between the stars are unimaginably huge, the time needed for interstellar travel is correspondingly large, and any system which may have a chance of overcoming this restriction requires what are to us impractical amounts of power.

The problem begins with Einstein's special theory of relativity, first published in 1905, when Einstein was working at the Swiss Patent Office in Bern. The special theory draws upon two firmly established scientific principles but comes up with another, which many scientists and non-scientists alike consider to be one of the weirdest notions in the whole of science.

The first of these principles comes from the work of Isaac Newton, who, in the 1680s, showed that the laws of physics are the same for any observers moving at a constant velocity relative

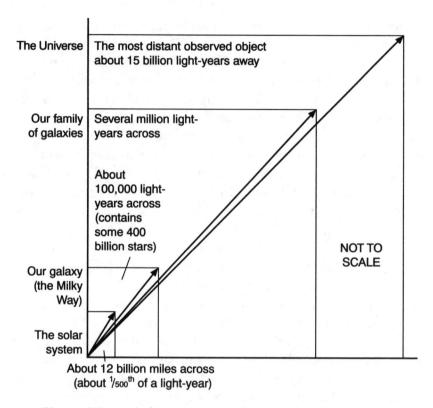

The Universe | The most distant observed object about 15 billion light-years away

Our family of galaxies | Several million light-years across

About 100,000 light-years across (contains some 400 billion stars)

NOT TO SCALE

Our galaxy (the Milky Way)

The solar system

About 12 billion miles across (about $\frac{1}{500}^{th}$ of a light-year)

Fig. 14: 'Size order' in astronomy: from our local solar system to the unobservable limits of the Universe.

to one another. So if a driver and passengers in a car travel alongside another car (or travel towards the other car), and both cars are moving at a constant velocity, the driver and passengers in each vehicle will each perceive the Universe to be behaving in the same way. This seems obvious, but it has important ramifications.

The second fact, arrived at more recently, is that the speed of light in a vacuum is always constant. This velocity is represented by the symbol 'c' and is equal to just over one billion kilometres per hour. But most importantly, this is true *irrespective* of the velocity of the observer.

According to common sense, if spaceship A is moving in one direction with a velocity of 0.75c, and spaceship B approaches from the opposite direction also travelling at 0.75c, their relative velocity would be 1.5*c*. But this is not actually the case. According to Einstein's equations, crews on each ship would see light from the other coming towards them, not at one and a half times the speed of light, but at just under 1*c* (0.96*c*, to be precise).

The astonishing consequence of this is that if *c* is constant, then space and time must be relative. In other words, if the crew aboard spaceship A or B are to see light arriving at a constant velocity, irrespective of their own velocity, they must *measure time differently* – so, as they travel faster, time slows down. Furthermore, because distance, time and speed are all interrelated, if time slows down, then the property of distance cannot be the same to observers travelling at different speeds. In other words, in this situation, if we change the time, then by logic, measurement must also change. The faster one travels, the shorter any given distance becomes – a metre will be a different length depending on the velocity of the observer, and will be shorter the faster the observer moves. Finally, the faster an observer moves, the more massive he becomes.

The end result of all this is that if it were possible for an observer to travel at the speed of light they would experience three things – time would slow to nothing; they would shrink to nothing; and their mass would be infinite!

And this is no mad theory or unproven hypothesis. Einstein's special theory of relativity has been proven to be true in many thousands of experiments conducted since 1905. But why, you might ask, did no one think of it before? Why didn't Newton realise what Einstein deduced? And most importantly, perhaps, how is it that when we travel along the motorway, Granny does not suddenly balloon to elephantine proportions, and our watches always appear to read the same time as those of people standing still or moving at different speeds?

The answer to all these questions lies with the magnitude of the speed. The reason we do not notice this effect every day of our lives is that we do not travel anywhere near fast enough. If he had considered the idea of travelling at speeds close to that of light, Isaac Newton may have deduced relativity in the seventeenth century; but even the value of the speed of light was not known then, so he may, I think, be forgiven.

An experiment conducted aboard a recent shuttle mission illustrated the consequences of special relativity and highlighted how minuscule the effect is at low speeds. Even travelling in orbit at the not unimpressive speed of 5 miles per second, clocks aboard the shuttle ticked less than one ten-millionth of a second slower than their counterparts on Earth.

The closest we can currently get to producing near-light speeds is not with massive objects such as spacecraft, but using single elementary particles. At CERN, the giant particle accelerator near Geneva, and at Fermilab near Chicago, sub-atomic particles are accelerated to near-light speeds routinely, and their masses are seen to increase precisely as Einstein's calculations predicted they would.

So, this law of Einstein's, this rule that nothing can travel at the speed of light, is irrefutable – it is simply a fact of life in our Universe. Consequently, the only possible ways by which we could ever one day hope to cross interstellar distances is either to travel at speeds which do not incur too many problems with Einstein's theory, but get us there eventually, or else we have to create some ingenious way of getting around the theory without breaking the rules – something humans have, in the past, shown themselves to be rather good at.

So, firstly let us consider the ways in which we might one day travel between the stars at sub-light speeds.

This possibility has been one of the staples of science-fiction for the best part of a century, and has occupied the thoughts of a growing number of scientists in recent decades. To most aeronautical engineers and space scientists it is the only conceivable way in which we or any other civilisation that may exist in the Universe could travel interstellar distances.

All conventional space propulsion systems (and by that I mean engines that do not use some exotic property of space itself, such as warping or wormholes), must work on the principle of Newton's third law of motion, which states that: 'For every action there is an equal and opposite reaction.' In this way a spaceship is no different from a jet aircraft or a powerboat. Material is expelled from the back of the craft and the craft moves forward – simple.

The difficulty for the interstellar explorer is not this system itself, but a question of magnitude. To travel to even the nearest star during a single human life-span, the vehicle would have to be driven by a very powerful device to provide the necessary acceleration.

The space vehicles we have developed so far all work by chemical propulsion – mixing together materials that react to produce energy which is squirted through a tiny hole at one end of the engine, pushing the craft forward. This differs little from the principle of jet propulsion.

The greatest energy requirement of any vehicle ever built has been the energy needed for it to escape the Earth's gravitational pull – to achieve what is called the escape velocity (for the Earth this is about 11km/sec). This is the energy needed by the Saturn V rockets that lifted payloads into Earth orbit, facilitating the first step to the Moon, or that needed to get the shuttle or the Ariane craft along with their payloads into orbit. Once in space, where gravitational forces are far weaker and where there is an almost complete vacuum, things get easier, and so all manoeuvres aboard the Apollo craft travelling to the Moon, for example, required only relatively small engines and thrusters that expelled hot gases from their exhausts and adjusted the course of the spaceship once it was in orbit. Without these, the capsules would have been entirely at the whim of the gravitational forces at work beyond the Earth's atmosphere.

This form of propulsion represents where we stand today. Some unmanned craft have used solar power or small nuclear reactors to generate energy to operate their on-board machinery, but chemical power remains the state of the art as far as propulsion systems are concerned. However, the next level of sophistication – and one which is hopefully not too far off – is some form of manned fission-powered spacecraft engine.

Nuclear fission is the power source used in terrestrial nuclear reactors and nuclear submarines, and was responsible for the tremendous energies unleashed in the earliest atomic bombs. When large unstable atomic nuclei are made to decay, they are said to undergo nuclear fission, and the result is energy. The value of this energy depends upon the mass of material undergoing fission, and can be calculated using perhaps the most famous equation in history (and again derived by Albert Einstein): $E = mc^2$, where m equals the mass of material and c is the speed of light.

Nuclear fission is the most powerful controllable energy source we have at our disposal today, and is responsible for a growing percentage of the energy that generates electricity to power our homes, offices and factories. Although it has its associated dangers, particularly the waste materials (primarily plutonium-239, which has a half-life* of 24,000 years), and always carries the risk of a catastrophic accident, it is a potent and very adaptable source of energy. But in any discussion of energy requirements for space missions, nuclear fission will only ever play a major role in short-hop, interplanetary travel because it could never provide the immense power required for practical interstellar travel.

Returning to our bubble analogy, nuclear fission could help the tiny creatures living in the bubble to get around inside their bubble, but it would be of very limited value in any effort they may make to reach another bubble across the ocean. And, if they were floating around somewhere near, say, Fiji, it would be almost totally useless in helping them reach America.

Even travel within our solar system would never be workaday using nuclear fission. The amount of fuel needed to power a spaceship would be so large it would leave little room for the crew or the cargo. Furthermore, as the accident at Chernobyl in the former Soviet Union has shown, nuclear power has potential hazards that would make space travel even more dangerous than it already is. Imagine the consequences if the ill-fated space shuttle *Challenger* had been ferrying nuclear material into orbit to power a large vessel under construction for a future Mars mission. Techniques may be developed to make such transportation into orbit safer, but most people would be uncomfortable with such a project.

A more powerful form of nuclear energy comes from a process

---

* Half-life is the time needed for a source of radioactive material to decay to half its original mass.

called nuclear fusion. Back in 1989 there was a brief flurry of excitement when two scientists, Martin Fleischmann and Stanley Pons, claimed they had devised a technique called 'cold fusion', which appeared to require nothing more than a pair of electrodes and some commonplace chemicals placed in a jar. Unfortunately, their experiments proved unrepeatable, and attention returned once more to the conventional attempts to develop controllable nuclear fusion.

Fusion is the mechanism by which the Sun or any star is powered. In the laboratory, the process involves fusing together small nuclei such as deuterium and tritium (which are heavy isotopes of hydrogen) to produce large amounts of energy.*

For almost fifty years scientists have been trying to develop practical nuclear fusion – so far with only limited success. Nuclear fusion has a great deal going for it. It is a relatively clean energy source, because it does not use dangerously radioactive elements such as the uranium-238 which is converted into plutonium-239 in modern fast-breeder reactors, and it has the potential to produce far more energy than nuclear fission. These are the plus points of the system; the downside has so far been the problem of containment and efficiency.

To bring about fusion, temperatures of around 10 million degrees centigrade are needed (the sort of temperatures produced in the core of the Sun), so that the positively charged nuclei can be forced to overcome their electrostatic repulsion. This fused material exists as a super-heated plasma which cannot be kept in any form of physical container. Furthermore, the energy needed to bring about fusion has so far been much greater than the energy return, which means the system currently shows negative efficiency.

Despite this admittedly serious drawback, nuclear physicists

---

* A heavy isotope is a version of an atom that has more than its usual complement of neutrons in its nucleus. The most common form of hydrogen has just one proton in its nucleus and no neutrons. The first heavy isotope of hydrogen, deuterium, has one proton and one neutron. The heaviest, tritium, has one proton and two neutrons.

hope to crack the problem in the near future and fusion is seen as the most likely way in which we could save the Earth's looming resource crisis. Sadly, however, even this energy source could only find very limited use for interstellar travellers from Earth, because once again, the amount of fusible material needed to achieve even a tiny percentage of light speed would be too great to make it viable.

It has been calculated that to accelerate a spaceship to just 5 per cent of the speed of light would require almost eight times its mass in fuel. And this is to accelerate just once. If the craft wanted to stop at its destination it would need to use more fuel, equivalent to almost four times the current mass of the ship. If we assume the outward voyage has used up half the fuel (which weighed four times the mass of the living quarters and cargo), a further four times the ship's mass would be needed again. So, one start and one stop would need [8 × 4 × the mass of the main body of the ship (excluding fuel)], or 32 times the mass of the living quarters and cargo.

One possible way round this problem is the idea of the fusion ram-jet. Although we think of space as being completely empty, in fact it contains hydrogen atoms distributed very finely between the stars and planets. It is theoretically possible therefore to construct a spacecraft which has large scoops attached to one end which would draw in the hydrogen atoms to use as fusible material.

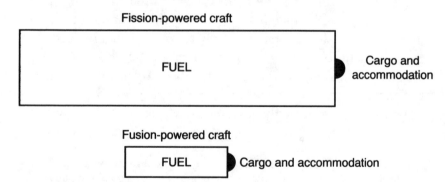

Fig. 15: A comparison of the size of craft needed using nuclear fission and nuclear fusion.

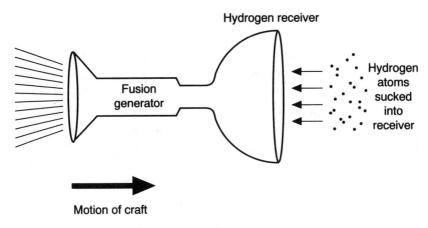

Fig. 16: The fusion ram-jet.

At first glance this design seems totally impractical, simply because the distribution of hydrogen is so fine – there would be insufficient material available in space to generate sufficient power. But if the craft is moving quickly enough it would behave like a person running through light rain, getting soaked because they are meeting the raindrops as they go, or similar to a giant sea mammal drawing in plankton, eating as it swims. Such a vehicle might look ungainly, but it would be able to power itself and travel at a respectable speed in terms of interplanetary travel.

Nuclear fusion, then, is viewed as the most likely way in which we will one day travel between the planets of our solar system on a routine basis – assuming problems of containment and power efficiency can be overcome. It is a practical fuel source for interplanetary travel (which would require speeds in the region of several hundred thousand kilometres per hour), allowing us to get to Mars in a few weeks. But, as we have seen, there is no comparison between interplanetary distances (travelling within the bubble) and interstellar distances (traversing the ocean). Using fusion power to achieve speeds of, say, 250,000kph, we could reach the outer limits of our solar system in about two years, but it would take us a thousand generations merely to complete a one-way trip to the nearest star, and the fuel requirements to maintain even this relatively trivial velocity for so long would alone make it totally impractical.

Apart from the more conventional ideas surrounding the use of atomic power – both fission and fusion – there have been several attempts in recent decades to devise rather more unusual systems to produce high speeds using nuclear technology (which is, after all, the most advanced energy source we have). One of these is the idea of using the power of nuclear explosions to thrust the craft forward.

Designers of a theoretical vehicle known as 'Orion' visualise using a store of thermonuclear warheads individually propelled from the back of the craft at the rate of one every three seconds. The explosion would produce a hot plasma that would then impact on what they describe as a 'pusher plate', propelling the spaceship forward. Although this would again offer the opportunity for exploration within our solar system (if there was some way to make such a device entirely safe), estimates of the efficiency of this system indicate that at best it could achieve a speed of just 3 per cent of the speed of light (25 million kph). To obtain this velocity a total of almost 300,000 one-ton bombs would be required, making it almost completely impractical.

More promising than any of these schemes could be the possibility of using exotic material known as antimatter.

All matter in our Universe is made of atoms, and these are composed of what are known as sub-atomic particles – neutrons and protons, which exist together in the nucleus of the atom, and electrons, which surround the nucleus. This much was understood early this century largely due to the work of such pioneers as Ernest Rutherford, James Chadwick, Max Planck and others. Another influential and innovative physicist of this era was Paul Dirac, who in 1929 predicted that all the known sub-atomic particles could have counterparts with opposite properties.* These became known as antiparticles.

Electrons are negatively charged; an antielectron (or positron) would have the same mass but would be positively charged. An

---

* In those days only protons and electrons were known. The neutron was not discovered until three years later, in 1932.

antiproton is negatively charged, and, like a proton, it exists in the nucleus of what could be thought of as anti-atoms. But what is most important for the future engineers and designers of interstellar engines is the fact that when matter and antimatter come into contact, they annihilate each other instantly and at the same time produce energy.

In Paul Dirac's day, antimatter was merely a theoretical concept, something that had popped out of the equations when he had combined the mathematics of quantum mechanics, electromagnetism and relativity. Because it is destroyed immediately it makes contact with matter, antimatter is not found naturally in our Universe, and until recently it could not be pinned down by experimenters. Today, we can manufacture small quantities of antimatter in a particle accelerator.

To make an antiproton, 'normal' protons are sent whirling around the accelerator ring, where they are accelerated in an intense magnetic field until they reach about 50 per cent of the speed of light. They are then allowed to collide with the nuclei of metal atoms. This process generates pairs of particles and antiparticles (along with X-rays and various other forms of energy). The antiprotons are then separated from the protons before they can interact and obliterate one another.

But how can such a rare material, which is so obviously difficult to handle, be used as a propellant? Obviously, to use this resource we would have to be able to control the annihilation of particles and antiparticles and to use the heat generated by the process to power a spacecraft. And indeed, a simple design for just such a system is already on the drawing-board. The idea is to fire a tiny quantity of antimatter into a hollow tungsten block which is filled with hydrogen. The particles are instantly annihilated, and the energy released heats up the tungsten block. Cold hydrogen is then squirted into the centre of the device where it is rapidly heated to about 3,000K and fired out of the engine, thrusting the vehicle forward.

The great advantage of an antimatter drive is that little fuel is needed to produce an effective acceleration. The great disadvantage lies in the difficulties involved in producing usable amounts

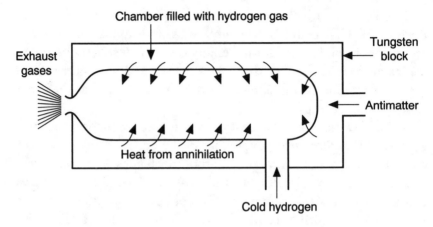

Fig. 17: An antimatter drive.

of the stuff. Currently only a sixty-millionth of the energy used in producing antimatter in the world's particle accelerators ends up as particles. For this reason its manufacture is incredibly inefficient, and its current market value is about $10,000,000,000,000,000 (ten thousand million million dollars) per gram.

With antimatter we also have the problem of containment. Like the super-hot plasma produced by nuclear fusion, special magnetic containment systems will have to be devised to prevent antimatter interacting with matter before it is needed.

None of these difficulties precludes its use by advanced civilisations – our future selves, or other civilisations elsewhere in the galaxy and beyond. We need only consider the development of our own technology to see that no technical problem is insurmountable. It was only shortly after the end of the First World War that Ernest Rutherford discovered that the nuclei of certain atoms could be made to disintegrate by bombardment. Within little more than a quarter of a century this work had led to the atomic bombing of Hiroshima and Nagasaki. And think of the difficulties natural philosophers of Newton's time would have had in imagining controlling the power of a Saturn V rocket or designing an engine for a jet aircraft.

Current technology may mean that antimatter is prohibitively expensive to produce, but these technical problems will almost certainly be overcome within twenty or thirty years. And if we look at the broader picture – the lifetimes of civilisations and the possible ages of advanced civilisations on other worlds – two or three decades becomes relatively meaningless. From the viewpoint of a human being reviewing history one thousand years hence, the period between the first atomic power stations and the mastery of antimatter engines will seem trivial. This may also be the case for other civilisations.

The option of antimatter propulsion systems offers hope that interstellar travel may be a possibility, but even if we do realise the full potential of this technology, we would still not be able to circumvent the natural laws of the Universe, and would remain limited to sub-light speeds. What this means for interstellar voyagers is that they will either have to accommodate the consequences of travelling at close to the speed of light (which will still be very slow for the purposes of colonising or visiting many worlds), or else travel even more slowly and take even longer to arrive anywhere outside their own solar systems.

If we imagine for a moment a journey to a star system which we believe could contain life, and which is 100 light-years from Earth, at $0.95c$ (that is, 95 per cent the speed of light), this would take 105 years to complete, one way. However, thanks to the peculiar effects of special relativity, the journey would only appear to take 105 years to those left back home. Because relative time slows as we travel faster, to the crew of the spaceship, this 105 years will be a little over 31 years.

Naturally, this is still far too long to be of practical use. Even if we were able to develop an efficient system of suspended animation, which kept the crew asleep and slowed their metabolisms dramatically, round-trips of 62 years (plus the time spent at the destination) would not be popular for a number of emotional, practical, ethical and commercial reasons. Like the problems offered by the development of antimatter drives, the technical difficulties associated with deep sleep or suspended animation will eventually be overcome, but there are other barriers to its use.

Crews sent out on such long trips might return home to find the political structure utterly transformed. The organisation or

institution that had funded the voyage and developed the tech-
nology behind it may no longer exist. Such a crew would almost
certainly find their relatives had died long since (as 2 × 105
years – over two centuries – would have elapsed on Earth, while
they themselves would have aged only 62 years), and almost
everything that was once familiar to them would be altered
beyond recognition. Even the difference three decades (one leg of
the above-mentioned journey) can produce is greater than one
might at first imagine. Think of the differences between the time
England last won the World Cup in 1966 and now. The fact that
the Beatles were in the charts and miniskirts and long hair were
popular is not so significant; after all, most of us can at least iden-
tify with these things, and fashions have a habit of repeating
themselves. More important would be the shifts in the attitudes of
society, the nature of institutions and the role and nature of gov-
ernments. Yet more significant still would be the technological
differences between the two eras. The equipment used by voy-
agers returning to Earth who had set off in 1966 would seem
antiquated in the extreme, even to us today. And this example,
remember, lies within a relatively tiny time-frame during the
latter half of the twentieth century. Technology is developing at
an exponential rate, and with it many once-sacred traditions of
society and the attitudes of individuals are also changing rapidly.
If this exponential change continues, in a few centuries from now,
the differences felt over a thirty-year time-frame will be consid-
erably more dramatic.

Beyond this is the problem of command structure and financing.
Journeys that would take anything longer than, say, a decade
could easily incur problems, because institutions and leaderships
change, governments rise and fall and priorities shift, making it
almost impossible to manage lengthy missions. Furthermore, any
endeavour has to see a return for the investor within a reasonable
time-frame – and certainly within the lifetime of the investor.
Problems would inevitably arise because long journeys could not
meet this condition, and would have to rely upon philanthropy.

The only other possible option for sub-light travel is a con-
cept that has for long been one of the staples of the science-

fiction writer – the notion of the 'space ark'. The idea of the space ark is to build a large spacecraft in which generations of travellers could live throughout a mission that may last hundreds or even thousands of years. Although this would require a vast spacecraft capable of sustaining a large crew and complement of passengers for perhaps millennia, it would not need to travel particularly fast. If a mission was designed to take 5,000 years, a distance of 250 light-years could be covered at just 5 per cent of light-speed (just over 40 million kph). Of course, the construction of such a vast craft capable of sustaining so many people is not even conceivable using current technology; but we may not even need to consider these difficulties, because even before the designer picks up his pen the concept has so many built-in drawbacks.

First, there is again the problem of time-scales. The only possible reason why any world government (for it would be unimaginable for a single state to attempt such a thing) would finance such a mission would be to escape global catastrophe. In other words, it would be an ark in the biblical sense, precipitated by a global cataclysm and seen as the only means of escape; a system by which a portion of the world's population might be packed into spacecraft which then headed off into the great unknown.

This all sounds terribly exciting and romantic, but what are the disadvantages, other than the unimaginable technical difficulties? How would the group sustain itself materially and psychologically?

We must assume that any civilisation capable of building a craft that could house a few thousand travellers would have long before overcome the technical considerations – methods of self-support, the growing of food and the recycling of water and food.* But, in some respects, the psychological considerations may present greater difficulties than the technical problems. The early generations of travellers aboard the ark would have no hope of

---

* For those readers interested in the technical considerations behind such an idea, I recommend taking a look at the science-fiction novel *Rendezvous With Rama*, by Arthur C. Clarke (published by Orbit).

seeing a new world, and would be kept going only by the knowledge that their distant offspring might make it to a distant planet. However, we need only look at the criminal disregard we humans show towards our environment here on Earth, and our lack of care for the health and quality of life of our descendants, to see that this motivation would be difficult to sustain.

But, more important even than the psychological problems the idea of the ark presents, the most persuasive argument against its use – other than as the last resort of a dying world – is the idea of the 'speed exponential curve'.

This is a simple graph which shows how our civilisation has developed techniques to travel exponentially faster as time goes on. During the first 100 millennia of human social progress, the highest speed we could reach was about 15kph – the pace of a sprinting hunter. This was more than doubled some 4,000 years ago with the domestication of the horse. Another doubling occurred towards the end of the nineteenth century thanks to the development of trains and motor vehicles; it was multiplied a further three or four times during the following fifty years with the invention of aircraft, and again with the advent of jets. Finally, spacecraft, built within the past forty years, have allowed a further quadrupling of the maximum speed at which humans can travel (to some 40,000kph). This change follows an exponential pattern, and if the curve is continued it can be seen that it should be possible for humankind to reach 5 per cent of the speed of light by the year 2140 and 10 per cent by the end of the twenty-second century.

Although this may sound like good news, for the ark people, it would completely destroy their *raison d'être*, because long before they arrive at their destination they would almost certainly be overtaken by the descendants of those they had left behind on Earth, scooting past in the latest spacecraft designed to travel many times faster than them.

A variation on the idea of the ark is the concept of long-term colonisation. The physicist Frank Tipler has lent his support to the notion that we may one day be capable of colonising the galaxy by 'planet-hopping'. He bases this idea on the way the

South Sea Islanders spread across the Pacific Ocean by island-hopping followed by consolidation. Using this model he believes there are two time factors to consider. First, the time needed to make an interstellar journey (t_1); the other, the time needed to establish a colony and to prepare for the next hop (t_2). Conservative estimates for the journeys would be in the 1,000–10,000 year range, and a reasonable period for colonising and consolidation would be in the region of one hundred generations.

Using this system, even travelling at a modest speed (he does, after all, allow interstellar journey times of 1,000–10,000 years), the human race could colonise the entire galaxy in a surprisingly short time, because the growth would be exponential. The first group would arrive on planet A after a long journey from Earth. This landing party would then consolidate and after perhaps one hundred generations they would have the resources to launch two new missions. These would land on planets B and C, and the process would begin again. Bearing in mind that the technology of the travellers should improve constantly, and there is no practical reason why knowledge should be lost, it is likely that the journey time (and perhaps the consolidation time) would be markedly reduced as they progressed. This would enable even more rapid galactic colonisation.

But even if we assume a constant time-frame for each step, the journey time and the colonisation time combined equal an average of 10,000 years, $t_1 + t_2 = 10,000$; assuming there are 1 billion suitable planets in an average galaxy, these could all be reached and colonised in under 1 million years (see Chapter 10).

It is a sobering thought that perhaps the Earth was once the site of colonisation, and that for whatever reason the colony did not prosper and the 'wave' of colonisation moved on, leaving us behind (see the next chapter, 'ET's Calling Card'). If this was the case, such a colonisation process might mean that all human life in our galaxy stems from a single mother planet, an original home of humanity which developed on a planet where life began to evolve earlier than it did elsewhere. Alternatively, it could be argued that such a colonisation process lies in our own future, and that Earth is the original, perhaps unique, home of *Homo sapiens* and that in the coming millennia we will spread throughout the galaxy and beyond.*

Looking at the options offered here for practical interstellar travel, the future looks unremarkable. Of course, assuming we survive, the human race will make the entire solar system its home within perhaps a century or two. We should be able to enjoy weekend trips to the Moon and holidays on Mars within the lifetimes of our grandchildren. Mining the asteroids should provide a practical source of minerals in the not-too-distant future, and even the further reaches of the system, the border territory of the distant planets, Uranus, Neptune and Pluto, will be thoroughly explored one day. Within a thousand years there will almost certainly be human beings living on every world in the solar system. This is in itself very exciting, but it does not hold out much hope for finding and communicating with alien life, especially intelligent life. For that we will have to venture to the stars, and to make that a practical option we need to get there far more quickly than we can visualise at present. Journey times to the stars would have to be brought down to a maximum of a few months rather than millennia. And there are, according to even the most adventurous theorists working in the realms of interstellar travel, only two possible ways in which this could be achieved.

Both of these ideas attempt to bend the rules of physics (since, by definition, they cannot be broken). They lie within the realms of exotic or borderline physics and remain entirely theoretical; neither are in any way practical for the foreseeable future. However, these concepts may harbour the seeds of methods which our distant descendants could use one day to travel thousands of light-years as easily as we travel across the Atlantic.

The first suggestion is the idea that a theoretical construct called a 'wormhole' might be one way to travel to the stars. The concept of the wormhole derives from a long-established and repeatedly verified piece of what is now viewed as traditional

---

* For those interested in this hypothesis, I recommend another science-fiction classic, Isaac Asimov's *Foundation* trilogy, and for a more detailed analysis, my own biography of the author – *Asimov: The Unauthorised Life*, published by Orion.

physics – Einstein's general theory of relativity. But building a picture of how a wormhole could exist, and perhaps act as a method by which interstellar travel was made possible, requires stretching the ideas of general relativity to the limit (some say *beyond* the limit).

Scientists have known for a long time that when a star has used almost all its available fuel it begins to die, and the way in which it dies depends upon its mass. If it is about three times the mass of our Sun or larger, it begins to shrink, setting up shock-waves which result in an enormous explosion – the most violent event since the Big Bang. This is called a supernova. But even then, because the sun was so large to begin with, some material is left at the centre of the supernova. This begins to collapse in upon itself again, forcing the matter to become ever more dense so that the incredibly strong forces holding sub-atomic particles together are overwhelmed and the star becomes a seething cauldron of fundamental matter and energy. This is a black hole, so-called because it is so massive and dense that even light cannot travel fast enough to escape its gravitational field.

Einstein's general theory of relativity, published in 1916, is an extension of the more limited theory of special relativity, which is only concerned with observers moving at a constant velocity. But after developing this theory, Einstein began to wonder what the situation would be for objects experiencing acceleration. He imagined a lift in a state of freefall, and a beam of light entering a hole in one wall. People in the lift would perceive the light travelling in a straight line. But to an observer outside the lift, the light would travel along a curved line. This bending of light, Einstein stated, was caused by the fact that the lift was experiencing acceleration. He went on to say that because gravity is a form of acceleration, light would be bent by it.

Until Einstein, physicists saw the Universe in three dimensions, with time as an extra quantity. In general relativity, time is a dimension, just like length, breadth or depth; and the modern view is that the Universe actually exists in four dimensions called 'space-time'.

The only way we can visualise a four-dimensional Universe is by representing it in three dimensions. Imagine a rubber sheet stretched flat. Now place a heavy ball in the middle – the sheet around the ball is misshapen. In a similar way, space-time distorts around a massive object such as a star. Roll a marble along the sheet near the heavy ball and it follows a curved path, just as light does near a star. A black hole is so massive and has such a powerful gravitational field that it curves space to the extent that within it lies what is called a 'singularity', a point at which the curvature of space-time becomes infinitely sharp and all the laws of physics break down. Wormholes, as theorised by a number of scientists (including the physicists who first postulated them – Kip Thorne and Michael Morris at CalTech in California), are created when two singularities 'find' each other and join up.

The reason wormholes are useful to interstellar travellers may be visualised if you think of one end of a wormhole existing near the Earth (for example, just on the rim of the solar system) and the other opening out at some far distant point elsewhere in the Universe, perhaps thousands of light-years from here. Because of the nature of curved space-time, wormholes offer a short-cut, by-passing the need to travel between point A (close to the Earth) and point B (the other mouth of the wormhole) using the conventional route. All that would be required for a journey of many light years would be a trip to the mouth of the wormhole (which may take a few weeks using atomic fusion generators or a few days using antimatter drives), and a corresponding journey at the other end.

I say 'all that would be required . . .', but the fact is, getting to the mouth of the wormhole – even if it was on the very rim of our solar system – is the very least of our problems. Although the idea of the wormhole cutting out the need to traverse normal space for perhaps many hundreds or thousands of light-years is obviously a very attractive one, the practical drawbacks are immense.

First, wormholes are still pure speculation. They are not disallowed by the known laws of the Universe, but neither are they certain to exist. But, assuming they do, they would probably be quite rare. The second problem is that it would be impossible to know which parts of the Universe a wormhole linked until it was used. Furthermore, if it were usable, it would offer only a very limited service, linking the starting point to one fixed destination. It would be a bit like having a motorway connecting London with some other mystery location with no junctions or turn-offs en route.

Even if we ignore this disadvantage, we have to consider the nature of the link, and from what we know of black holes, a 'natural' wormhole would offer a very bumpy ride indeed. The inside of a black hole is probably the most inhospitable place in the Universe; the gravitational forces at work there would instantly decompose any material object into a soup of fundamental particles and energy – and even if these forces could be resisted, once within the grip of the black hole, there is no escape. So, the idea of using a wormhole created by joining two black holes at two different points in the Universe does not seem very practical. The only way they could be used would be if there are certain types of black hole somewhere in the Universe that do allow passage, but these might be very difficult to find.

A possible way around this difficulty is the idea of 'white holes'. As the name implies, these theoretical objects are the very opposite of black holes. According to the theorists, rather than absorbing matter and energy, white holes are thought to act as perfect emitters, or 'cosmic gushers'. So, if a black hole and a white hole were joined, they could act as a one-way wormhole and circumvent the logistical problems of emerging from the 'far end' of a wormhole. Unfortunately, detailed mathematical analysis of this scenario has shown that such a system would be unstable, and that the white hole would decay rapidly, making the passage of a spaceship impossible.

An alternative to the use of natural wormholes, and a great hope for the enthusiasts, are man-made wormholes.

Since wormholes were first postulated by Thorne and Morris in a paper published in the *American Journal of Physics* in 1987,[*] hundreds of theoretical physicists around the world have studied the

---

[*] Kip Thorne created the idea of wormholes for his friend, Carl Sagan. At the time, Sagan was working on *Contact*, his novel about human communication with aliens, and had written asking Thorne to propose a scientifically plausible method by which aliens could travel interstellar distances within a practical time-frame.

idea and added new concepts to the rostrum of theories surrounding these far-out mathematical constructs. And one of the conclusions that has been speculated upon by those working in the area is that, in order to construct a workable wormhole, a set of strict conditions would have to be met. These include the obvious fact that the construction of the wormhole must be consistent with general relativity, and that the gravitational 'tidal forces' within the wormhole be kept to a minimum. They also stipulate the shape to which the wormhole must conform and the mass of material needed to create it. Unfortunately, the mathematics shows that in order to construct an artificial wormhole, material known as 'exotic matter' is needed, which has the bizarre property of negative mass.

Although wormhole enthusiasts insist that such a nonsensical idea can be realised within the laws of physics, most scientists currently dismiss the notion. If they are right, it would appear that wormholes could never be manufactured, no matter how advanced a civilisation becomes, because the laws of physics cannot be broken. If they are wrong and the wormhole supporters are correct, then exotic matter has to be found and manipulated by the civilisation before the wormhole could be built and used.

The second possible way to stretch or bend the laws of physics is perhaps even more outlandish than the idea of man-made wormholes. This is the concept of the warp-drive, or travel through 'hyperspace'. Like the notion of the ark, this hypothetical means of interstellar travel has captured the imaginations of science-fiction writers and fans alike, and is one of the key ideas in the TV series *Star Trek*.

The warp-drive has been visualised as a device that could enable a craft to circumvent the impracticalities of sub-light speed travel and the nuisance of having to work within the laws of physics. Even so, it stretches dramatically some cherished concepts of physics.

Another name for warping, and one which perhaps describes it more accurately, would be 'surfing'. This is because it is based upon the principle of manipulating space-time itself so that the space vehicle moves on a 'wave'. The spacecraft would have the ability to alter space-time so that it expands behind the craft and contracts in front of it. This means that even though the craft is itself moving relatively slowly, the departure point would be 'pushed' back a vast distance and the destination 'drawn' nearer.

This sounds like a cheat, but again it is a possibility within the rules of general relativity. The difficulty is not that it is theoretically impossible, just practically impossible in the foreseeable future because of the absurdly large amounts of energy required to make it work.

Observation of our Sun shows that its mass curves space-time so that it bends light by just one thousandth of a degree. For a spacecraft to utilise the expansion and contraction of space-time itself it would have to distort the space-time continuum far more than this. In some respects the vehicle would have to behave a little like a tiny black hole. Using this as a basis for calculating the energy requirements, the result sounds depressingly familiar. To make a black hole the size of a typical spaceship, say a disc 20 metres in diameter, we would need a mass of about 30,000 Earths compacted into the space. Expressed in terms of energy, this would be about equal to the entire energy output of the Sun during its lifetime.

So what are we to conclude from this survey of propulsion systems? It is quite clearly nonsense to suggest that we will never get to the stars – less than two centuries ago, before the first test of Stevenson's Rocket, a serious claim was made by an eminent physician that the human body could not tolerate speeds above 30kph, and that passengers would be crushed aboard the train. Furthermore, unless there is a dramatic U-turn in the attitude of humans, a reversal of our innate drive to discover, we will continue to confront and overcome the technical problems facing us, whether it is on Earth or in the field of space exploration. Space research is expensive, and despite its natural glamour it is not popular with a large proportion of people, who see it as a waste of money and have never realised the great benefits we have already gained from little more than three decades of space exploration. Given these facts, development, at least in the near future, will be slow.

It is quite likely that we will probe into the furthest recesses of our solar system, but as this summary highlights, the effort required to reach for the stars is in a league of its own. But eventually the problems will be overcome. Perhaps the earliest missions to other stars will be slow, unmanned efforts, the interstellar equivalent of an ox and cart hobbling to its destination and overtaken many times by more advanced vehicles.

Whichever way it is done, one day humans will stand on the surface of a planet orbiting another sun and feel the heat of that alien star. Beyond that will come a time when we will develop the means to generate the huge energies required to power craft to travel at close to light-speed, and perhaps even to devise ways to circumvent the restrictions laid upon us by laws unravelled by Einstein and others. When that day dawns, true interstellar travel will be with us.

But it is also quite possible that, long before then, beings from other worlds, perhaps only one tiny step ahead of us, will get here first. Maybe they have even already been . . . and gone.

9

ET's Calling Card

'If God did not exist, it would be necessary to invent him'

VOLTAIRE

Picture the scene: a giant UFO hovers over the desert. Laser beams extend from under the craft and manoeuvre huge slabs of stone into position, as astonished nomads look on from atop their camels or from vantage points far away from where – it seems – the gods themselves are wielding their incredible power. Later, when the craft has left, three huge pyramids remain in the sand, each the largest manufactured objects on the planet's surface. And for generations, the confused and amazed tribespeople recount the events in their legends and their sacred texts, describing the alien craft as fiery messengers of the gods and their pilots as angels in golden chariots.

According to some, this scene could really have been enacted some 5,000 years ago at Giza, the site of the Great Pyramid, and similarly at different times elsewhere in the world – at Stonehenge in England, on Easter Island and at Machu Picchu in Peru. Exactly why some advanced civilisation would build these things, or help the local people to do so, remains one of the great unanswered questions of paranormal investigation, but theories abound, ranging from their use as 'cosmic transmitters' to their functioning as 'gateways' to the stars.

Some enthusiasts of the ancient-astronaut theory even go so far as to suggest that the human race itself was 'seeded' by alien visitors, who arrived on the planet hundreds of thousands of years ago and gave evolution a nudge – as depicted in the opening scenes of Arthur C. Clarke's *2001: A Space Odyssey*.

For many people, the question of whether or not the Earth has been visited in the distant past is beyond doubt – they see the answer in the many stone edifices and ancient monuments scattered about the planet. But is this notion little more than another conspiracy theory, a concept that should be left within the realms of science-fiction and fantasy?

Occultists and mystics of Victorian times wove intricate webs of pseudo-science to try to 'explain' how alien cultures were in contact with the people of Earth, and how cosmic communities were both morally and scientifically far in advance of us – beings who wished us to join the great happy family of interstellar travellers. The Theosophy Society was one such group, whose leaders purported to be in contact with the souls of superior beings who lived on different 'levels of existence' on other worlds. But with the popularising of science early in the twentieth century, when science actually became a lot more exciting than some of the rather jaded ideas of the occultists and spiritualists, interest in the Theosophists and other mystics declined. Suddenly, people were a lot more interested in hard facts – the startling worlds offered by science-fiction writers who extrapolated upon such concepts as relativity, rocket engineering and life on Mars – concepts which were, for the most part, completely incomprehensible to the average person, but nonetheless totally compelling.

For some time, little further thought went into the idea of ancient astronauts or alien visitors walking the Earth during the infancy of civilisation, until a Swiss researcher and writer named Erich von Daniken startled the world with wild and often bizarre claims described in a series of best-selling books published in the early 1970s.

In his first book, *Chariots of the Gods?* (first published in 1969), Von Daniken revived the almost forgotten idea that the great ancient monuments of the world – the pyramids at Giza, Stonehenge, the Inca and Maya temples, the giant figures of Easter Island and others – could have been built by visiting aliens (or at least their construction influenced by benign extraterrestrials).

This was a marvellously media-friendly idea. It shocked and offended orthodox religion with headlines asking 'Was God An Astronaut?'; it captured the feeling of the time – the post-hippy depression and the so-called dawning of the 'Age of Aquarius'; and generated a sense of hope that, in some strange way, massaged our collective ego, giving the insecure a sense of greater self-importance. But gradually, Von Daniken found critics even within the ranks of those who wanted to believe.

Soon after publication of *Chariots of the Gods?* an American television programme called *Nova* exposed a collection of ideas contained in the book. Von Daniken had claimed that pieces of ancient pottery depicting what were interpreted as UFOs pointed to evidence of ancient visitation, but the programme researchers showed these to be crudely faked, and even managed to track down the artisan who had produced them.

Von Daniken also claimed that because the Maya knew the Venusian year lasted 584 days, this was clear proof that they had been visited by aliens who imparted this fact to them. But the true Venusian year is actually 225 days, a fact realised only recently, when the complex orbital behaviour of Venus was unravelled using what astronomers call the Doppler shift.

After the success of his books, Von Daniken appears to have gone off the rails completely. He invested in a hotel in Switzerland but was convicted of fraud and embezzlement meriting a three-year prison sentence. He has recently returned to writing and has published a new book, *Return of the Gods*, in which he revisits the themes covered in his multimillion-selling collection of the 1970s. This is obviously a canny move, considering the end-of-millennium obsession with all things spiritual or occult, but despite the fantasy and the frauds, is there at least a kernel of truth – or even half-truth – in Von Daniken's pronouncements?

First, let us look at the claims of Von Daniken and others,

because not surprisingly these are not entirely uniform, and different enthusiasts propose different schemes for our alternative global history, some of which are more far-out than others.

The most straightforward theory is that an advanced race of travellers came to Earth some time around 3,000 BC and passed on some of their knowledge to the ancient Egyptians, who incorporated this new understanding into the construction of the pyramids. Enthusiasts of this theory point to the fact that the engineers of the Nile region appear to have suddenly learned how to build pyramids properly around 4,600 years ago. The ruined pyramid of Meidum at Al-Faiyum, about 100 miles from Giza, in which cedar beams were used in some of the chambers, is, it is claimed, evidence of a rapid learning process. Believers also emphasise the fact that many constructions of so-called primitive cultures living in different parts of the world are remarkably similar, which suggests that they had a common teacher.

Of particular importance in this line of reasoning is the use of the number pi, or 3.142. According to archaeologists, the importance of this number was discovered by Archimedes, who lived in Greece during the third century before Christ. But the ratio of the height of the Great Pyramid to the perimeter of its base is almost exactly twice pi (to within an accuracy of a few inches), and the Pyramid of the Sun in Teotihuacan, Mexico, demonstrates a ratio between its height and the perimeter of its base of precisely four times the figure of pi. The probability of this figure being used by chance *twice* is extremely small. Supporters of the alien-influence theory claim that the Egyptians and the ancient peoples of Mexico were contacted by groups of extraterrestrial visitors, who perhaps lived among them and taught them mathematics, engineering and architecture, and that this is reflected in the mathematical integrity of the major pyramid sites.

Of course this begs the question: why?

According to some enthusiasts, the motivation was purely selfish – the alien visitors wanted to construct a device that would serve their purposes. What this 'device' could have been is again open to wide interpretation. Some claim that the pyramids, Stonehenge and other monuments serve different functions. The pyramids, they argue, could be 'gateways' to the stars or perhaps 'markers' for the aliens' colleagues during their interstellar travels.

Presumably, a marker or cosmic transmitter is some device left behind by alien visitors that could act as a beacon for future missions, or to transmit information gathered on Earth. What a 'gateway' may be is hard to pinpoint. It could be anything from a domesticated wormhole used for interstellar travel to some other advanced device developed by a technologically superior race and used to transport matter and bodies across space.

As we saw in the last chapter, the notion of interstellar travel using wormholes has become fashionable in recent years, and the idea that an advanced civilisation utilised such a device based at the Great Pyramid has inspired at least one mediocre film, *Stargate*. Unfortunately, there is not a scrap of scientific evidence to support such a concept, and, sadly for us, the ancient Egyptians left very few records of what they were doing or how they did things.

Slightly more sober suggestions include the idea that the newly-educated Egyptians actually designed and built the pyramids after 'the gods' had left, and that their construction was a form of homage to their teachers. A variation upon this theme is the idea that the ancients were attempting to create a massive 'model' of the heavens by arranging the pyramids in the form of certain stars they observed in the sky – presumably (or so the enthusiasts insist), because the star system at the centre of this plan (in the constellation Orion, according to some) was the home of the alien visitors.

These then are variations of the simple version of the ancient-astronaut theory – the idea that human societies were educated by a group of aliens (or perhaps several groups) who visited here at one time or another, perhaps during the era preceding the construction of the pyramids in Egypt and Stonehenge in England. Why the aliens were here in the first place is, of course, an open question, with a wide selection of answers that range from the idea that an alien craft crash-landed and left the occupants stranded, to the notion that the visit was deliberately planned for an unknown purpose. An alternative is the idea that an alien race stopped by during an 'ark mission'.

A variation upon the central idea of alien visitors is the theory that advanced beings arrived here in our deep past, and that our evolution was prompted by extraterrestrials or even that humankind was 'seeded' by them.

For some, the whole idea of evolution by natural selection presents problems. They point to the fact that, according to some theorists, there has been insufficient time for the massive changes that have occurred in our past, changes that have allowed us to evolve from a shrew-like mammal some 50 million years ago to *Homo sapiens* and modern humankind. Indeed, this question remains a matter for hot debate among some evolutionary biologists and is not entirely explicable, but it strikes most rational scientists as unnecessary to jump to remarkable conclusions about alien influence in order to explain what is really only a glitch in evolutionary thinking and a matter that will almost certainly soon be explained by orthodox means.

The need for some form of extraordinary origin for humanity is an understandable weakness, but just because it is a nice idea by no means makes it fact. We know so little about the way life can evolve in the Universe that it would be rash to say that life does not form easily and therefore has to be seeded. It is also an idea that smacks of pseudo-religion, little more than a variation of the creation myth. If we accept the idea that life on other worlds is plentiful, and, further, that advanced civilisations other than our own have evolved and continue to evolve elsewhere, then it is also logically consistent to say that those civilisations could have begun their development long before us. But it is an enormous jump (even from this point of reasoning) to believe that one such advanced race came here perhaps millions of years ago and set us on the road to humanity.

So what is the evidence for alien visitation in ancient times?

Because all physical clues have long since disappeared, there is almost nothing to support the idea. Furthermore, unless humanity is part of some unimaginably massive cosmic scheme, then there would appear to be no real reason why an alien intelligence would interfere in a natural process. Of course, it may be that we are indeed part of some such gigantic plan, but if that is so, we will almost certainly never know about it, and as there is no clear evidence for such a faith-based concept it is a rather pointless avenue to pursue.

Supporters of the idea that there have been advanced human civilisations that existed and thrived on Earth in ancient times suggest that these communities may have sprung from alien visitors who arrived and then remained here. Enthusiasts point to the many and diverse legends incorporating advanced but lost civilisations in our own deep past – legends of Atlantis and Mu, for example, which are supposed to have maintained sophisticated technologies tens of millennia before the first clear signs of civilisation as delineated by conventional archaeology. Again, sadly, the evidence is lost or so vague as to be open to varied interpretation.

In terms of being able to offer any form of supporting evidence, the best alternative scenario to orthodox history and prehistory is the idea that alien visitors came here around 5–6,000 years ago and influenced the development of already established civilisations before disappearing again. But despite being slightly more prosaic than the full-blown 'alien seeding of humanity' idea, it still offers little in the way of hard evidence. All evidence for this idea is circumstantial and centres around interpretation of a collection of ancient mysteries and, in recent years, some rather far-fetched speculations based upon findings from unmanned probes sent out into our solar system.

Alternative archaeologists (as I will call the group of investigators who believe in the ancient-astronaut theory) have gathered together a set of findings which they claim point to the presence of alien interference at various stages in our past. These break down into several distinct categories.

First, there is the 'evidence' found in the ancient writings and records of various cultures from around the world and from different eras. Second are interpretations of the purpose of ancient monoliths and theories about how these were constructed; and third are artefacts which, it is claimed, may have been left behind by alien visitors or represent development of an ancient culture far beyond that accepted by orthodox archaeology. Finally, images sent back to Earth from recent missions to the other planets have, say the enthusiasts, given us a glimpse of a wider cosmos which, if interpreted correctly, could offer us a window into what they believe to be our 'true' history.

The first type of evidence (that from ancient records) offers the most colourful and initially convincing material to support the ancient-astronaut theory. This is due partly to the interesting consistency between cultures and eras even when, according to orthodox archaeology, there was no physical or material communication between them. We have already seen how the value for pi was almost certainly known and used by at least two ancient peoples who lived at different times and had no known contact. The alternative archaeologists can offer many examples of highly suggestive ancient texts which also possess a similar uniformity. Many of these raise a number of questions about their origins. Take, for example, the ancient Indian text, the *Mahabharata*. At several points the authors describe what are called 'Virmanas', a word meaning 'flying machines'. In the text, the Virmanas are sometimes described as flying vehicles used for military purposes, and are often piloted by Indian gods. One of these is given the name the 'Agneya weapon', and it appears in one particularly striking passage, which recounts:

A blazing missile possessed of the radiance of smokeless fire was discharged. A thick gloom suddenly encompassed the hosts. All points of the compass were suddenly enveloped in darkness. Evil bearing winds began to blow. Clouds reared into the higher air, showering blood. The very elements seemed confused. The sun appeared to spin round. The world, scorched by the heat of that weapon, seemed to be in a fever. Elephants scorched by the energy of that weapon ran in terror, seeking protection from its terrible force. The very water being heated, the creatures who live in the water seemed to burn.

At first glance this might appear to describe something like a nuclear weapon, which has led enthusiasts to conclude that these Virmanas must be of extraterrestrial origin. Statements such as 'clouds reared into the higher air' immediately conjure up a vision of a mushroom cloud, but they could equally well describe the plume of a volcano or a forest fire.

Even more emotive is the statement 'the very elements seemed confused'. Anyone with a little knowledge of science is aware that elements are transmuted during nuclear processes, but the people who composed the *Mahabharata* would have known very little modern chemistry. To them, matter was composed of 'elements' akin to the Greek notion of the four elements – fire, earth, air and water. In fact, this passage (and many like it from other ancient texts) is entirely open to interpretation. It is possible that this account could have originally described a natural phenomenon such as a volcanic eruption, but the tale has been adulterated by numerous reinterpretations by one writer after another.

A more interesting case is the following passage, taken from a translated hieroglyphic text based upon an eye-witness account from the Egyptian Pharaoh Thutmose III, around 3,500 years ago:

In the year 22 [meaning the twenty-second year of Thutmose III's reign], of the third month of winter, sixth hour of the day, the scribes of the House of Life found that there was a circle of fire coming from the sky. It had no head. From its mouth came a breath that stank. One rod [about sixteen feet] long was its body and a rod wide, and it was noiseless. And the hearts of the scribes became terrified and confused, and they laid themselves flat on their bellies. They reported to the Pharaoh . . . Now after some days had gone by, behold these things became more numerous than ever. They shone more than the brightness of the sun, and extended to the limits of the four supporters of the heavens. Dominating in the sky was the station of these fire circles.

Again, according to supporters of the ancient-visitors theory, this describes a typical UFO sighting not so very different from any number of reports made this century, but sceptics would suggest that the account describes some form of natural phenomenon such as ball lightning (a compressed bundle of charged particles which behaves like lightning) or perhaps a tornado.

Another form of record left by early civilisations is pictorial representation of what really mattered to them as individuals or as a society. In this way the ancient Egyptians and the ancient peoples of South America recorded the legends of their gods. One of the most striking pieces of pictorial evidence to support the claims of the ancient-astronaut enthusiasts is a drawing discovered in the ancient Mayan temple at Palenque in Mexico. It shows a human figure seated in what looks astonishingly like a modern space capsule. The figure is squeezed into a small space jammed with levers and what could be interpreted as control panels, and coming from the rear of the contraption appears to be a flume of smoke and fire, not unlike the vapours expelled from a NASA rocket.

This is not the only picture from the ancient world that depicts what could be interpreted as space technology. According to some alternative archaeologists, primitive man seems to be obsessed with space-suited figures. One drawing found in cave dwellings in Val Camonica, northern Italy, depicts figures which may be interpreted as astronauts. They are dressed in large suits and what look like helmets and visors. Another interpretation may be that the drawings were actually showing nothing more exotic than the hunting gear worn by primitive people during a period now recognised as a mini Ice Age.

Similar drawings have been found at ancient American Indian sites in North America, in Uzbekistan and at Tassili in the Sahara. Naturally, not all of these could be interpreted as members of a tribe wearing cold-weather hunting clothing, but reasonable alternatives could include protective clothing used to withstand desert storms, or even ceremonial robes. Alternatively, the artists could have represented important members of their society as larger-than-life figures, and only in the Space Age are we able (and willing) to reinterpret these figures as space-suited aliens.

A variation upon this theme is found in the records preserved by ancient cultures as verbal accounts. A striking example comes from the West African Dogon tribe, who knew of the existence of a star called Sirius B – a star which can only be seen with the aid of a powerful telescope and was first photographed only as recently as 1970. What is even more impressive is that the tribesmen knew this star was a member of what modern astronomers call a binary star system (a star that orbits another); in this case the much brighter Sirius. Astonishingly, the Dogon tribe even knew the duration of the star's orbit – around fifty years.

The author Robert Temple, who has written a book about the discovery, *The Sirius Mystery*, believes the tribespeople learned about Sirius B from the ancient Egyptians some 3,000 years ago. Others even suggest that this information was passed on to the pharaohs by alien visitors. However, astronomers claim this knowledge is nothing more than coincidence, and point to the fact that a high percentage of star systems are binary and that the figure of fifty years was a lucky guess.

For the alternative archaeologist, the most important monument surviving from the ancient world is undoubtedly the Great Pyramid at Giza. Ever since early archaeologists returned to Europe with tales of these gigantic constructions in the sands, the pyramids have sparked the imaginations of paranormal investigators and occultists around the world.

One of the original Seven Wonders of the Ancient World (and the only one remaining today), the Great Pyramid is a truly awesome object. Erected around 2500 BC, it is the largest pyramid in the world, measuring 230 metres on each side of its base and originally measuring 147 metres high. To put that into perspective, you could fit four blocks of Fifth Avenue along one edge of the base, and St Paul's, Westminster Abbey, St Peter's, Notre Dame and York Minster could all fit comfortably inside the pyramid at the same time. It contains upwards of 1 million blocks of stone, each weighing about 2.5 tonnes. According to Napoleon Bonaparte (who was admittedly prone to exaggeration), there is enough stone in the Great Pyramid to build a wall ten feet high and a foot thick all the way around France.

These are undoubtedly impressive statistics, and according to believers, such an edifice could not possibly have been built by ancient Egyptians without the help of an advanced technology. Yet this is simply not true. Orthodox archaeologists have constructed a detailed description of how tens of thousands of slaves were used to drag the stones from boats which had brought them from quarries in the Lower Nile. They have plotted the route of roads specially designed and built to transport the stones, and have shown how Egyptian engineers had the mathematical and engineering skills to construct a building of such specifications and which utilised accurate numerical relationships between the dimensions of the length of edges, the height to the apex and the square base. As Dr Trevor Watkins of Edinburgh University has said of this argument: 'Just because the Egyptians and others did not leave behind books on BSc Engineering in Pyramids doesn't mean they didn't have them.'

Others are less circumspect. Michael Wright of the Science Museum, London, has said: 'The claim that aliens helped build the wonders of the ancient world is absolute bunkum. People can do anything they want to do with the right tools, persistence and patience.'[1]

There are many theories which attempt to describe the purpose of the pyramids, and even those who firmly believe in alien visitation are in disagreement over the matter. However, with the help of some advanced technology of our own, some of the mysteries of the pyramids will perhaps be solved in the near future.

In 1993, a German team lead by a researcher named Rudolf Gatenbrink used a specially designed miniature robot to video 60 metres of previously undiscovered tunnel inside the Great Pyramid. At the end of the tunnel the robot found a 20cm by 20cm door with metal brackets. Using special laser equipment attached to the robot, Gatenbrink was able to determine that a hidden chamber lay behind the door. This discovery has quite naturally excited researchers working on the theory of ancient alien visitors, and hopes are running high that when this door is eventually opened, clear documentary proof will be discovered to support their ideas. Rudolf Gatenbrink is that rare beast – a serious scientist and a friend of the writer Erich von Daniken, and few people have become more excited about this new work than the Swiss enthusiast for the ancient-astronaut theory.

'No one can deny the new secrets in the Great Pyramid,' he has said. 'They will back my theory that the extraterrestrials were here and that they created intelligence by a method of artificial mutation, and they will return.'[2]

What Von Daniken is suggesting here is that the Earth was not a natural repository for the elements that created intelligent life – that evolution by natural selection was not enough. For some reason, aliens not only came here, but *had* to come here if life was to evolve into an intelligent civilised form. So, not only does such a theory require belief in alien life and the ready manipulation of technology advanced enough to allow interstellar travel, but an acceptance that, for some reason, life here needed a shot in the arm from a more advanced civilisation to enable us to reach our present level of physical and social development. For many scientists, it is this last concept that is inconsistent with fundamental scientific principles.

To support their claims, the alternative archaeologists point out that an American clairvoyant named Edgar Cayce prophesied that vital records would be discovered in a hidden chamber in the Great Pyramid before the end of the century. But again, this is hardly evidence; it might prove to be little more than a lucky guess (even if these documents are found). Most scientists rightly view such occult claims as purely circumstantial, and many orthodox archaeologists are actively hostile towards these suggestions, viewing the researchers into the paranormal as nothing more than obsessive cranks.

Another ancient monolith worthy of comparison with the Great Pyramid is Stonehenge, and among occultists and alternative archaeologists it is the subject of outlandish theories about its construction, every bit as bizarre as those used to explain the construction and purpose of the pyramids. Lying eight miles north of Salisbury in Wiltshire, the earliest construction at the site began a few hundred years before the beginnings of the Great Pyramid at Giza – about 2800 BC. But unlike the Great Pyramid, the Stonehenge site evolved over a period of almost 1,800 years. Conventional archaeologists have identified four different phases of construction, with Period I beginning around 2800 BC and Period IV ending about 1100 BC.

Theories concerning the use of the site and the way such an edifice could have been constructed by primitive tribespeople are varied and plentiful. Again, enthusiasts of the ancient-astronaut theory suggest that Stonehenge is one of many sites situated on ley lines – hypothetical lines of 'force' or natural energy which intersect at key points (such as Stonehenge). What these ley lines are actually for, or why they should be considered important, is still a mystery to the investigators of the occult. To the scientist they are little more than imaginary 'lines in the sand', and the grid-work of lines that may be drawn linking the various stone circles dotted around Europe signify precisely nothing.

Again, the key factor in the importance of Stonehenge for the alternative archaeologist is the notion that they could never have been built by primitive people some three millennia before Christ. Yet, once again, conventional archaeology can offer a clear picture of how it was done. Several scholarly works have appeared in recent years describing in intricate detail the methods employed by the ancient Britons, and the techniques they employed using the materials readily available at that time.[3]

The constructions produced during Period I were very primitive – little more than a crude circle built around a burial ground. In Period II (sometime around 2100 BC), people of what has been called the 'Beaker Culture' built an earthwork approach road, now called the Avenue, which led to the original bank surrounding the monument of Period I. They also set up within the earlier ring a double circle of menhirs (large, rough-hewn standing stones) which came originally from the Preseli Mountains of south-west Wales.

The inner ring with the huge stone lintels that give the monument its characteristic features was built quite late, almost 1,000 years after the first rough-hewn stone circle. Known as the sarsen stones, these were transported nearly twenty miles from the Marlborough Downs, dressed to shape with stone hammers and jointed together. The skill employed by the builders is peerless within European prehistory, and may only be compared to that employed by the stonemasons of Egypt, but it is at best insulting to the human spirit of endeavour to suggest that the construction of Stonehenge or the pyramids was only possible with the help of alien visitors.

On the other side of the world from Stonehenge and the pyramids of Giza lies another wonder drawn into the cannon of evidence offered by the alternative archaeologists. This is the set of markings found on the Plain of Nazca in Peru.

From the air, these markings can be seen to make up a vast network of patterns. Most of the markings are straight lines that bisect and interlink with other shapes – sometimes those of birds or animals stretching for miles across the plain. It is thought that, as with the construction of Stonehenge, the markings were created over a long period, probably sometime between AD 600 and 1200. But the astonishing thing about them, and the reason for their value to the alternative archaeologist, is that they only take on distinguishable patterns when observed from the air, and were only discovered when modern explorers first flew over the site during the early days of aviation.

As the discovery became widely known, investigators of the paranormal and believers in the ancient-astronaut theory began to speculate that the markings on the Nazca Plain were landing sites for visiting UFOs. Erich von Daniken has even speculated that some of these odd shapes could be parking bays for aircraft or space vehicles.[4]

To the orthodox archaeologists, the markings are nothing more supernatural than Inca roads. Yet, at first, this is hard to reconcile with the fact that these 'roads' seem to lead nowhere; and the purveyors of paranormal explanations decried this theory as more far-fetched than their own. But the conventional argument might be correct. It has been pointed out only recently that if many of our cities were mysteriously destroyed, and a thousand years later our motorway system was observed from the air, it would appear that many roads also lead nowhere.

Whatever the purpose of the markings, one thing has been clarified – there is no need to apply extraterrestrial explanations to describe how the markings were constructed. The archaeologist Maria Reiche has shown that it is actually relatively easy to expose the darker rock which lies just beneath the surface layer of the plain. And in 1992 David Browne, an archaeologist who works with the Royal Commission for Ancient Monuments, showed that the lines cut into the rock of the Nazca Plain could have been produced by 10,000 people working for just a single decade. Furthermore, because the lines are known to have been produced over several centuries, it would have required far fewer workers to have produced them, perhaps no more than 1,000 at any given time. His explanation for their construction is that they could be thought of as 'spirit paths' − roads that lead to vast open spaces used for religious ceremonies. And this would also tally with the explanations of conventional archaeology, which document the huge efforts ancient peoples were willing to go to in the construction of religious symbols or places of worship.

If this interpretation is correct, then the pyramids, Stonehenge and the lines of the Nazca Plain may be thought of in the same light as Westminster Abbey or St Peter's in Rome − expressions of devotion to God, or gods, for which no expense or effort would be spared.

As well as the great structures that have survived the ravages of time and the evolution of human society, there are the many smaller artefacts that signify the characteristics of a civilisation, material objects that show the level of technology and the preoccupations of a society.

The Holy Grail for alternative archaeologists, proof of their theory, would be something like a skull found to be 5,000 years old but possessing crowned teeth, or a variation upon a machine-gun carbon-dated to three or four millennia before Christ. Sadly for the enthusiasts, nothing of that nature has been found, but to be fair, the likelihood of unearthing such things so many thousands of years later would be remote, even if there had been a few hundred alien visitors possessing advanced technology based here. Yet the believers in the idea that we have been visited by aliens can point to one item which offers the best form of evidence so far discovered.

In 1936, archaeologists working in Iraq found what is now called the 'Baghdad battery' − a device consisting of all the components of a working cell, minus the battery acid. However, this was not a discarded battery from a local store, but is estimated to be 2,000 years old. When researchers made an exact replica of the device, and used fruit juice as the battery acid, it produced half a volt of electricity. Although this would be insufficient to power very much, alternative archaeologists have suggested that the solution may have been much stronger, or instead that the system was used to coat objects using the modern technique of electrolysis. Either way, this discovery points towards a highly sophisticated technology operating two millennia ago. This could, of course, have been a technology created by humans, but what is odd about it is that there appear to be no other related discoveries. One would expect a society which had developed the battery to have used it in some way, and over a period of time to have developed associated technologies.

However, there are a number of arguments against this. First, after so long a time it is reasonable to suggest that any other devices that may have been created by the same people would have been lost, and that the Baghdad battery was a lucky find. Second, there is the idea that it may have been a one-off invention created by a 'Leonardo da Vinci' of the time. Third is the argument that some societies have indeed produced great advances in one direction, but remained relatively backward in others. The ancient indigenous peoples of South America showed remarkable architectural skills, but are thought never to have developed the wheel.

The final source of 'evidence' for the alternative archaeologists is material from the space missions of the past few decades (particularly unmanned interplanetary missions) which, they claim, has thrown up anomalies that they insist are inexplicable using conventional arguments.

Of these, the two most significant are what have become known as the Martian face and the Martian pyramids. Both of these were first noticed in photographs taken on the Viking 2 mission in September 1976.

From the only photograph available from NASA, the feature known as the Martian face (to be found in the Cydonia region of the planet) does indeed look like a human face.* It is thought to be about one mile long and about 1,500 feet high. According to some researchers, this structure was made by intelligent beings as a signal to us that we are not alone in the Universe – the theory being that by the time we are advanced enough to see this object on the surface of Mars, we would be ready to accept that there is (or at least was) life elsewhere in the Universe that created such sculptures in the distant past.

The 'pyramids' are different from those found in Egypt. They are far bigger – something like a kilometre in height – and seem to be five-sided rather than four. Some seriously suggest that they are arranged in precise geometric patterns and exhibit profound mathematical arrangements, in a similar way to known relationships observed in the dimensions of our own pyramids.

Although the entire concept of artificially produced monuments on the surface of Mars may seem far-fetched, the idea does have the support of some serious researchers. A former NASA consultant, Richard Hoagland, believes the pyramids were made to precise configurations by an alien intelligence. Another NASA physicist, Dr Brian O'Leary, who was originally a sceptic, has recently begun to concede that there might be something odd about the images.[5] And Carl Sagan once commented that the objects 'warrant . . . a careful look'.[6]

Without further evidence from more photographs and the long-awaited manned Mars mission, the true nature of the Martian face and the 'pyramids' will not be known, although the NASA have said that the Orbital Surveyor will pass over the sites. The chances are that the objects are merely optical illusions. It is a recognised psychological faculty of human beings to find patterns from chaotic and semi-chaotic data, and both the face and the pyramids could be nothing more than an effect produced by our need to stamp the familiar on the unfamiliar.

There is plenty of superficial evidence to convince those who

---

* It has been suggested that there was a second photograph taken from a different angle which does not show a human face, but this has not been released by NASA.

desperately want to believe in life on other planets and our part in a greater cosmic scheme, but none of it satisfies the rigorous demands of a sceptical scientific community, and it is right that scientists should view these matters dispassionately and with a healthy degree of objectivity.

However, as I hope this book clearly illustrates, there is nothing supernatural about the idea that the Universe is teeming with life, and there is nothing illogical or impossible about the idea that we could have been visited by highly advanced alien beings in our remote past. But the simple fact is that recourse to such a theory is quite unnecessary – the pyramids and Stonehenge could have been built using indigenous talent driven by religious fervour; the ancient texts could be describing natural phenomena; and the Martian face is almost certainly nothing more exciting than a striking rock formation. But until we can prove these things beyond doubt, it might be wiser to keep one eye open to extreme possibility within our otherwise orthodox view of the Universe.

10

INDEPENDENCE DAY

'There are more things in heaven and earth, Horatio,
Than are dreamt of in your philosophy'

WILLIAM SHAKESPEARE, *HAMLET*, I. V. 166–7

Suppose it were all true. Just for a moment, suspend your disbe-lief and contemplate what would happen if an alien craft landed in Trafalgar Square.

For many the matter is already settled: for them, aliens are here, and have been visiting us for millennia. For UFO enthusiasts it is not a question of 'if' or 'how', or even 'why', but 'when'.

It is now fifty years since the UFO phenomenon became a global media subject, and the flying saucer is now an icon of the modern world. Whether you believe the images we have all seen are dust-bin lids or craft from another planet, they are here to stay, and they will become further absorbed into the human collective experience. UFOs will never entirely disappear from our minds because the only way the image can change into a 'truth' is by the discovery of advanced alien beings who travel through interstellar space. This may happen, but the reverse never will – humans will never be able to say for certain that such beings definitely do *not* exist.

Every commentator on the subject of life on other worlds draws a line in the sand at some point. Each expresses a point of view. For some, the entire notion of any form of life beyond Earth is doubtful. Another large body of thinkers have reached the con-clusion that there is life in abundance but that none of it has reached the level of intelligent, self-aware beings.

As I'm sure you are by now aware, I totally disagree with this view and believe that if life originates on any world it is inevitable that this life-form will evolve and progress. There will be casualties and set-backs, self-destruction and evolutionary culs-de-sac – as there have been on our own world – but life would find a way.

There are others who have what I consider to be rather odd views about life on other worlds. The American physicist Frank Tipler is of the idiosyncratic opinion that there cannot be intelligent life in the Universe because if there was they would have destroyed us by now. In his scenario, any advanced technology should have long ago produced robot explorers that would have pervaded the entire galaxy, including Earth. Because there is no evidence this has happened, he believes there cannot be intelligent life elsewhere.*

This is a statement even more ridiculous than Enrico Fermi's flippant remark so fondly brandished by the extreme sceptic, and is even more woolly in its determination. For me, Tipler's statement demonstrates the worst aspect of scientific thinking. He refuses to accept any arguments against his view, and is dogmatic to the point where one wonders what has precipitated such a belief – does it derive from some twisted humanism or some childhood trauma? It is analogous to medieval thinking, the cherished, anthropocentric view that we are special. Rival thinkers have pointed out that aliens may have outgrown the idea of galactic domination, but Tipler merely responds by saying that he is basing his theory upon the way he perceives all life to behave on Earth.

In fact, Tipler and others of the school who believe 'alien beings cannot be there because we haven't encountered them' leave so much unexplained. They ignore the possibility that we have not yet contacted extraterrestrials because no race has reached a point where it can communicate across the galaxy. Most importantly, they ignore the fact that the Earth is a tiny insignificant dot in a vast ocean of empty space, positioned in a

---

* This concept is what Tipler claims to be the logical consequence of the planet-hopping model described in Chapter 8.

relatively out-of-the-way suburb of the galaxy and easily over-looked.

So, what other reasonable explanations remain?

First, we should at least pay lip-service to the idea that some UFO sightings are genuine encounters with alien technology. In purely emotional terms, this solution is probably the most desirable (so long as the alien visitors are friendly!). For centuries there have been countless reports of sightings and personal contact with what witnesses believe to be alien beings. The number of these claims has grown enormously in recent decades and mushroomed since the earliest media hype surrounding the phenomenon immediately following the Second World War.* Almost all of the several million reported cases can be explained away as illusions, weather effects, lies, misinterpretation, wishful thinking and psychological disturbance, but there are several incidents which still defy explanation. This does not mean that they are genuine encounters with advanced beings from another world, but that explanations have not yet been found to cover them.

For me, the simple fact that interstellar travel is so remarkably demanding and will be difficult for us to achieve in the coming millennia means that a civilisation would have needed to reach a very advanced stage of development in order to master convenient and therefore *rapid* travel around the Universe, shrinking travel times between the stars to hours or minutes as described in, among many others, *Star Trek* and *Star Wars*. But, as I have explained at some length elsewhere in this book, this could have already been achieved by alien beings who began their development a few thousand years before us.

However, this is not the entire problem. It is perfectly reasonable to believe that aliens have developed interstellar travel, but I

---

* This is in itself an interesting coincidence. Was the timing of the sudden global interest associated in some way with the aftermath of the Second World War? Perhaps people needed to replace the drama of world war with some other phenomenon, or maybe it was linked with Cold War paranoia. The fact that the media explosion surrounding UFOs began in the United States could be linked to something in the collective American psyche, or it could just be that the US had the most advanced electronic media at the time. This debate deserves a book in itself.

and many others do not accept that we are being visited on a regular basis. This is simply because, as far as we know (and that is admittedly an assumption based upon very limited knowledge), there is no real reason for them to do so. An advanced culture would require reasons to do things, just as we do. The idea that advanced aliens are snatching humans and animals to conduct experiments upon them, or, according to some reports, to have sex with them, is ridiculous, and those who believe such things are either mentally disturbed or have simply not thought it through. Where are the reasons?

The idea that we are witnessing a cover-up, a conspiracy between our governments and alien beings, is equally unbelievable. And the reason is the same – why? TV programmes like *The X-Files* and *Dark Skies* are wonderful entertainment, but they are no substitute for real life, or even logic, come to that. Such dramas as abduction by aliens and covert activity with shadowy organisations are far too unsubtle for what would have to be highly advanced beings. If they can travel across interstellar space they would have no need for such blunt instruments.

In my view, the idea that aliens are visiting is a very appealing one, but it is counter-logical. I think it is possible that aliens who have developed sufficiently advanced technology to travel freely around the Universe may have dropped by on occasion, they may even have stayed for a while long ago, but the idea that they are interfering in our development is completely without foundation.

In a multitude of different ways, science-fiction writers, and in particular the creators of *Star Trek*, have been remarkably accurate with some of their ideas and predictions. This is nowhere more apparent than in some of the sociological and psychological precepts they employ, the motivations and drives they weave into their plots. A very good example of this is the idea of *non-intervention*, and this brings me to another possible reason why there is no clear evidence for contact.

Would an advanced civilisation, one which has mastered interstellar travel and lived through the experiences of growing up in the Milky Way, really want to make contact with beings like us?

We are a cruel and savage species, but these characteristics are also intimately linked with our success. It is quite possible that our determination, our drive, our inquisitiveness, are simply the negative aspects of the same force that allows us to commit the foulest of deeds. But, if this is true, it is also almost certainly a universal pattern. Perhaps any animal that evolves into a technological species possessing self-awareness and self-motivation can do so only if they also have the psychological make-up to fight for their supremacy.

As we evolve emotionally and as a society, we are trying (with varied degrees of success) to accommodate our more negative instincts, to liberalise our thinking, to overcome some of the animal instincts that could destroy us now we have the hardware to do so. This does not mean we will ever or could ever completely eliminate our aggressive instincts (and it would not be a good thing to do so anyway), but we are gradually learning to control these characteristics.

With this in mind, how would we be perceived by an alien race? Would the equation balance? Would the works of Picasso, Dostoyevsky, Darwin and Buñuel negate Belsen, Bosnia and the Spanish Inquisition? Does a mother's love counterbalance the work of Jack the Ripper?

There is every reason to believe that an advanced alien race would have gone through similar dilemmas, catastrophes and triumphs to us. They too may have had their seasons of blood, but would they really want to wallow in the mire again? Perhaps J. G. Ballard had it right when, after Apollo 11 landed on the Moon, he said, 'If I was a Martian I'd start running now.'

There may be an entire spectrum of alien cultures in our galaxy, ranging from the most peace-loving to the most violent and war-like, but I can see little reason why any of them would want to become involved in our lives at the present stage of our development. And if there are such races in large numbers, then perhaps they have something analogous to a Federation of Worlds, in which the intervention of trouble-makers would be prevented by the majority. Equally, those with only peaceful intentions would not make themselves known to us until we had reached a technological stage where we might become a nuisance. If this is true, they might make contact deliberately when we develop our own means of interstellar travel.

This may all sound like another conspiracy theory, but on a galactic scale rather than one involving the grubby antics of our political leaders. Indeed, it is mere conjecture, but it might offer a possible reason for the fact that we have not yet made contact with any other intelligent beings – they're deliberately avoiding us.

But what might happen if one day in the not-too-distant future we awoke to find that we had made contact with an alien intelligence?

The answer to this would naturally depend upon what form this first contact takes. Perhaps the least dramatic (but still life-changing) scenario would be the detection of a signal which could be proven to be from an artificial source. This may come from our 'local' environment (say, within a fifty light-year radius of earth), which would mean that the senders are likely to still be around and communication of a sort could be established, albeit very slowly. If the signal is found to come from any further away, it would have been sent long ago and communication would prove difficult, if feasible at all.

The alternative is the possibility that a signal could be detected from a distant star positioned thousands of light-years away. This would mean one of two things: either the senders had become extinct, or they have evolved technology beyond our imaginings and may well be on their way to our sector of the galaxy. If a signal was received from so far away constructive communication would be almost impossible, as each message would take thousands of years to reach its destination.

Another possibility is the idea that an alien race will detect one of our signals. As I mentioned earlier, we have been advertising our presence as a technological civilisation for some seventy-five years (since the earliest days of radio). Therefore any civilisation situated within a 75-light-year radius of us would have a chance of detecting leakages of signals from Earth or deliberate messages we have broadcast. Some commentators have suggested that we are playing with fire letting ETs know we are here, but this fear has little logical foundation because any intelligence capable of visiting us to wreak havoc and eat us alive would have known of our presence long ago, and may even have been observing our development.

Another possible means of contact could be through the discovery of an artefact, either theirs or ours.

I mentioned the Martian 'face' in the previous chapter. If in the unlikely event that our return to Mars proved that this is indeed an artificially-produced image, then we would have to radically re-evaluate our theories concerning the evolution of our civilisation, our place in the Universe and even the history of our planet. But even if this were indeed found to be the staggering truth, there is very little we could do about it.

Obviously a manned mission to Mars would be mounted as quickly as possible. Hopefully an international effort would be made and we could be on the Martian surface within a few years, but the beings who produced such a thing may now be long extinct. Alternatively they could have moved on thousands or even millions of years ago. Perhaps, as I intimated in the previous chapter, such beings merely passed through at some time in our long-distant past.

An alternative to this is the possibility that ET finds one of our probes – Pioneers 10 and 11 or one of the Voyagers. In the coming decades more missions will be sent to the outer planets and these probes will be allowed to drift into interstellar space. Naturally, they will take hundreds of thousands of years to reach the nearest stars, assuming they survive long enough, but an advanced alien civilisation capable of travelling freely around the cosmos could pick them up long before then. Pioneer 10 is still transmitting radio signals (although we are no longer listening to it), so it would be detectable if the alien beings had versatile equipment that could trace something as quaint as radio signals.

Whatever the form of contact, the workers at the International Academy of Astronautics in the USA have already begun preparing for such an event, and have produced a document called the *Declaration of Principles Concerning Activities Following the Detection of Extraterrestrial Intelligence*. This is an eminently sensible piece of work which outlines the steps that should be taken if one day we do make contact with an alien civilisation. Constructed along the same lines as the *Treaty on Principles Governing the Activities of States in the Exploration and Use of Outer Space, Including the Moon and Other Celestial Bodies*, drawn up during the early days of space travel, this document quite correctly places the United Nations and the international scientific community at the centre of any procedure. Included in the declaration is the suggestion that the nations of Earth co-operate in the formulation of a response to an alien signal or landing, that no single nation tries to make unilateral contact, that a thorough investigation into the nature of the contact is made before the public are informed and that SETI, in collaboration with politicians via the UN, should spearhead the investigation and continued communication. It also rightly points out that if the contact is made via a signal received from a distant star, then the response should be carefully judged and thought through before transmission because there is no need to rush; after all, the message could take centuries to reach its destination.*

However, as exciting as these ideas obviously are, the really big question is: if aliens just turned up here one day, what would this mean for humankind?

Naturally, this all depends upon the intentions of the visitors. Some believe that such a contact has already been made, and cite the Roswell crash as a visit that went wrong. But now, over fifty years after Roswell, even some hard-core UFO enthusiasts are beginning to believe the incident was a fake. What is needed is a clear-cut, unambiguous contact, something straight out of *The Day the Earth Stood Still*.

If an alien race were capable of coming here then they would be highly advanced technologically, and we would be no match for them if they decided they needed our world for some purpose: the *Independence Day* scenario is an entertaining one, but hardly realistic. Even if we assume that advanced beings would be so totally amoral as the aliens in the film, and were capable of arriving here in such large numbers without being detected until the last

---

* For more information about the *Declaration*, try the Academy's website or the 'Leonardo' e-mail address: *isast@sfsu.edu*.

minute, technology as advanced as theirs would certainly not be vulnerable to a computer virus. But that's Hollywood. In reality, alien beings may not be interested in our world simply because there are intelligent beings here – why go to all the effort of wiping out an entire civilisation if interstellar travel is no problem?

Clearly, in the final analysis, if an aggressive and war-like race turn up in orbit one day, it would almost certainly be the end for humanity and there would be no hope of an *Independence Day*-type final reel. But there is also little to fear from this idea. It is far more likely that at some stage in the future, aliens will contact us when they feel we are ready. And what this will mean for us and our future is one of the most interesting questions in the field of extraterrestrial investigation. In the event of friendly 'physical' contact, the key areas most dramatically affected will be religion, science, politics and our response as individuals; our reaction to a shift of race image.

No major religion precludes the possibility of life on other planets, but some take a naïve anthropocentric stance. Christians believe the discovery of life on other worlds would not conflict with biblical teaching but merely reinforce the idea that God has created a diverse Universe filled with different life-forms. Even Evangelical Christians have little problem with the concept. When informed of the discovery of what may be Martian fossils in meteorite ALH84001, evangelist Keith Wills was reported in one newspaper as saying: 'It's all supposition at the moment, but I've no problem with the idea that there might be a simple form of life on Mars. I believe that God created the world in seven days. This doesn't make any difference to what Genesis says, that God created the world and put life on it and created human beings in his own image, with soul, spirit and body. If there was life on other planets it wouldn't be human life.'[1]

For the followers of Islam there is also no real conflict, although interestingly they insist that human beings were created from water and that there must be water on an alien world for there to be life – a rare example, perhaps, of reaching a scientifically correct conclusion from faith-based premises.

Yet despite these confident words, the arrival on Earth of

advanced alien beings would be the greatest shock imaginable to anyone with strong religious convictions. All religions are based on dogma, supported by legends and dependent on the belief that we are in some way special. Religion will be the first and perhaps the most easily damaged aspect of human culture when the big day comes.

But what could replace religion if it was so radically shaken up? It has been said that if humankind did not have religion, we would believe in anything. What would the arrival of a more advanced race mean for our natural religious impulses? If we had irrefutable proof that the galaxy is awash with intelligent life, that a Federation-style community of worlds existed, how could this really be reconciled with our quaint ideas pontificated about so confidently by the faithful? Would they be swept aside? Would a new era of enlightenment be ushered in?

Let us hope so. After all, what has organised religion really done for humanity? It has inspired some great artists who have produced many exquisite pieces of art in the name of God, some sublime pieces of music, a few worthwhile works of religious literature crafted for the greater glory of the Lord, but do these counterbalance all the wars that have been sparked off by religious bigotry? Does it make up for the torture inflicted by the Inquisition?

Many believe that the loss of orthodox religion could only do us good. Perhaps a more advanced race will show us where we have been going wrong, highlight just how bigoted and puerile we have been.

Science could also be given a tremendous boost. It would be wrong to assume that any advanced civilisation out there would just arrive on our doorstep laden with gifts – a hyper-space drive here, a cure for cancer there – but if such beings exist and they feel we are ready for contact, then we will almost certainly learn a great deal from them. Whether or not their own rules allow for such a leap in our development (or indeed if they felt inclined to assist), is another matter, and many believe we have some way to go before we could be accepted and allowed to join the party.

There is, however, another aspect to this point of contact.

Different areas of science do not develop in a vacuum – all science inter-relates. In order for our species to benefit from the knowledge of a superior civilisation, we would have to assimilate a broad spectrum of discoveries. There is no point, for example, in our being told how a warp drive works if we do not have the technology to build the rest of the craft – the navigational know-how, the materials technology, the means to build systems to protect the flesh-and-blood pilots against the inconceivable G-forces, the communications devices – the list is almost endless. By the same token, what good would it be for us to learn how to produce perfect artificial limbs if we didn't have the surgical techniques to graft them to our bodies? What use would a compact disc have been to a citizen of the seventeenth century?

This is one of the reasons for the application of a policy of non-interference, and would almost certainly govern the behaviour of friendly aliens if they ever decided to make contact. For this reason there probably never would be a landing in Trafalgar Square. A gradual unfolding of information, a 'soft' contact involving messages of increasing complexity and revelation is far more likely.

This would also cushion the blow for the politicians and leaders of the Earth, men and women who may be the most vulnerable to the shock of contact. The leaders of the world, we should recall, are soaked each day with a sense of untainted self-importance, and the more powerful they are, the more exaggerated is the delusion. How would the President of the United States feel if he were suddenly made to realise that he did not have the top job after all?

But it has been argued that this problem will impinge upon all of us in one way or another, for we are all convinced of our own self-importance. Despite the discoveries of science and the overwhelmingly clear fact that faith in some abstract construct is not what keeps us and society running, but that our lives depend upon science, the laity of all religions still place humankind at the pinnacle of creation and still believe we are watched over by a benevolent God. Five hundred years of scientific advancement from Copernicus to Stephen Hawking has actually affected the thinking of the vast majority of people very little. Just look at the way seemingly sane individuals follow astrology, surround themselves with superstitions and believe in 'luck'. Such medieval indoctrination is difficult to shift and lies just below the surface of everday life. An event such as an alien landing will shake our ingrained feeling of self-importance to its foundations.

Or will it? Could other factors be more influential?

Until recently, the notion of life on other planets was bunched in with a belief in witches, ghosts and life after death. Now, interest and research into the question of extraterrestrial intelligence is mainstream, almost conventional. This has been helped by news of possible Martian fossils, the discovery of planets orbiting other suns, and new ideas about how simple life can survive in the most extreme conditions. But perhaps of equal importance has been a psychological shift created by the world of entertainment.

For centuries writers have considered the idea of extraterrestrial life, and there have been many fluctuations in public interest in the subject. As I have discussed, Jules Verne, H. G. Wells, the science-fiction writers of the Golden Age of the 1940s and the B-movies of the 1950s all generated and nurtured huge interest in the idea of alien life and how we may interact with it. But, until recently, this remained purely an interest of what we would now call the 'anorak brigade'. Today things are very different. *Independence Day* is one of the biggest-grossing films of all time. On television, one of the biggest successes of recent years has been *The X-Files*. On the Internet, the (second) most popular search subject is anything linked with extraterrestrial life.

All of this must have a dramatic affect on our intellectual and emotional responses. Naturally, no one can say if there will ever be contact with alien intelligence, let alone how that may come about, but I would suspect that we are more prepared for such a thing than we have ever been before. And, with tongues only partly in cheek, some have suggested that all this is another conspiracy – that we are being slowly and carefully prepared for the big news.

Proof of extraterrestrial intelligence, however it is found, will be the biggest story of all time, but unless contact is sustained, unless we make physical contact and become intimately involved with other races, even this news will fade into the background.

My personal feeling is that the shock generated by the most important news in human existence will rock our world and its societies to their core. But some will be more prepared than others. Those with open minds and the ability to accept new ideas, to discard dogma and prejudice, will embrace such a paradigm shift willingly and happily. Hopefully we will see a new and very different world emerge from the afterglow; one that may become a better place because of it.

NOTES

1 LIFE ON MARS

1 Donald Brownlee, 'Ancient Life on Mars?', *Science*, Vol. 273, 16 August 1996, p. 864.
2 Leon Jaroff, 'Life on Mars', *Time*, 19 August 1996, p. 44.
3 Ibid, p. 48.
4 See Editorial in the *New Scientist*, 17 August 1996, p. 3; and W. Wayt Gibbs and Corey S. Powell, 'Bugs in the Data? The Controversy Over Martian Life is Just Beginning', *Scientific American*, October 1996, p. 12.
5 Robert Clayton, 'Did Martians Land in Antarctica?', *New Scientist*, 17 August 1996, pp. 4–5.
6 Gibbs and Powell, ibid, p. 13.
7 Roger Highfield, 'Organic Material In Martian Meteorite "Came From Earth"', *Daily Telegraph*, 16 January 1988.
8 Roger Highfield, 'I Found Life On Mars Seven Years Ago, Says Briton', *Daily Telegraph*, 1 November 1996.

2 WHAT IS LIFE?

1 Kevin Warwick, *March of the Machines* (London: Century, 1997), p. 19.
2 Primo Levi, *The Periodic Table* (London: Abacus, 1997), pp. 225–33.

3 Lee Smolin, *The Life of the Cosmos* (London: Weidenfeld & Nicolson, 1997). See also David Concar, 'Thank Heavens for Black Holes', *New Scientist*, 24 May 1997, pp. 38–41.

4 WHAT ARE THE CHANCES?

1 Marcus Chown, 'Seeds, Soup and the Meaning of Life', *New Scientist*, 17 August 1996, p. 6.
2 Justin Mullins, 'Other Worlds, Other Lives', *New Scientist*, 17 August 1996, p. 10.

5 SIGNALS FROM BEYOND

1 As quoted in Frank Drake and Dava Sobel, *Is Anyone Out There?: The Scientific Search for Extraterrestrial Intelligence* (London: Souvenir Press, 1993), p. 31.

6 THE GREAT PLANET HUNT

1 Robert Naeye, 'The Strange New Planetary Zoo', *Astronomy*, April 1997, p. 42.
2 Ibid, p. 46.
3 Ibid, p. 49.

7 THE CHASE

1 As quoted in Robert Zubrin, *The Case for Mars: The Plan to Settle the Red Planet and Why We Must* (London: Simon & Schuster, 1996).
2 Quoted by David S. F. Portree in *Astronomy*, April 1997, p. 104.
3 See John S. Lewis, *Mining the Sky: Untold Riches from the Asteroids, Comets and Planets* (Boston, MA: Addison-Wesley, 1996).

9 ET's CALLING CARD

1 Quoted in FOCUS, *Relics of the Paranormal*, December 1996.
2 Sarah Moran, 'The Truth is in the Pyramids', *Uri Geller's Encounters*, December 1996, p. 33.
3 See, for example, John North, *Stonehenge* (London: Harper-Collins, 1996).
4 Erich von Daniken, *Chariots of the Gods?* (London: Souvenir Press, 1969).
5 Jenny Randles and Peter Hough, *Encyclopaedia of the Unexplained* (London: Headline, 1995).
6 John and Anne Spencer, *The Encyclopaedia of the World's Greatest Unsolved Mysteries* (London: Headline, 1995).

10 INDEPENDENCE DAY

1 *Daily Telegraph*, 23 August 1996.

INDEX

Page numbers in *italic* refer to the Figures.

ABOUT THE AUTHOR

Michael White is a former science writer for *GQ* magazine. In a previous incarnation he was a member of the Thompson Twins pop group and a science lecturer before becoming a full-time writer in 1991. He is the author of a dozen books, including the international bestsellers *Stephen Hawking: A Life in Science* and *Isaac Newton: The Last Sorcerer*. He has also written biographies of Albert Einstein, Charles Darwin, Isaac Asimov, and John Lennon. *Weird Science*, his most recent work, explores theories of the paranormal.